SOME ASPECTS of the OPTIMAL CONTROL of DISTRIBUTED PARAMETER SYSTEMS

J. L. LIONS
Université de Paris and I. R. I. A.

SOCIETY for INDUSTRIAL and APPLIED MATHEMATICS
PHILADELPHIA, PENNSYLVANIA 19103

Printed for the Society for Industrial and Applied Mathematics by
J. W. Arrowsmith Ltd., Bristol 3, England

Contents

Preface

This paper corresponds to a set of ten lectures given at the National Science Foundation Regional Conference on Control Theory, August 23–27, 1971 at the University of Maryland, College Park, Maryland.

Actually some of the material presented in the lectures is not reproduced here, either because it would have increased too much the length of this (already long) paper (for example, controllability and stability) or because this material is in a form which is not yet ready for publication (for example, multi-criteria problems for distributed parameter systems[1]). On the other hand, we have included here some technicalities not given in the lectures.

In Chapter 1, after recalling some standard results of convex analysis, we give some examples of "variational inequalities" arising in mechanics.[2]

Chapter 2 gives some results already given, for instance, in Lions [1] together with new results (§ 7, § 8, § 9) related to controls depending only on time. The "dual" aspect for filtering theory is given in Bensoussan and Lions [1].

In Chapter 3, we study a nonlinear (distributed) system arising in biochemistry (cf. J. P. Kernevez [1], Kernevez and Thomas [1]). We give a nonlinear case where one can prove an existence and uniqueness theorem for the optimal control and we give some indications on the optimal control of systems governed by variational inequalities of stationary type (a problem dealt with in collaboration with G. Duvaut); we mention problems of optimal control of systems governed by variational inequalities of evolution which arise in biochemistry (Kernevez [2]). We conclude Chapter 3 by a remark of L. Tartar.

Chapter 4 gives existence theorems for geometrical problems of optimization: optimal positions of sensors, optimal design, etc.

Chapter 5 gives some indications on the possible applications of boundary layers in singular perturbations and how one can use these layers for simplifying the problem.

Chapter 6 gives a short summary of some numerical techniques; one has to consider this brief review as a (very) preliminary report, since much work is now under way in a number of places.

For an apparently almost exhaustive bibliography up to 1969 we refer to A. C. Robinson [1]. For other aspects of the theory, consult the books of Bensoussan [1],

---

[1] Papers on this topic by A. Bensoussan, J. P. Yvon and the author are in preparation.

[2] A systematic presentation of applications in mechanics and in physics is given in the book by Duvaut and Lions [1].

Butkovsky [1], the survey paper of P. K. C. Wang [1] and the abstracts of the IFAC Symposium on the Control of Distributed Parameter Systems, Banff, June, 1971.

The field of applications of the theory of optimal control of distributed parameter systems is extremely wide and, although a large number of (interesting) questions remain open, the whole subject is expanding very rapidly.

The author wishes to thank Professor Aziz for his help during the preparation of this paper.

CHAPTER 1

Partial Differential Inequalities and Unilateral Problems (Variational Inequalities)

1. Minimization of convex functionals. We recall here a number of well-known results in convex analysis. Let us consider a functional

$$v \to J(v)$$

defined on $U_{ad} \subset U$ and with values in $\mathbb{R}$; we suppose that:

(1.1) U is a reflexive Banach space[1] on $\mathbb{R}$;

(1.2) U_{ad} is a *closed convex* set in U;[2]

(1.3) J is *convex* and *continuous* on U_{ad}.

We consider the problem:

$$\inf_{v \in U_{ad}} J(v). \tag{1.4}$$

EXISTENCE THEOREM. *If we assume that*

$$J(v) \to +\infty \quad \textit{if} \quad \|v\| \to \infty, \qquad v \in U_{ad},\text{[3]} \tag{1.5}$$

then there exists $u \in U_{ad}$ *such that*

$$J(u) \leqq J(v) \quad \textit{for all} \quad v \in U_{ad}. \tag{1.6}$$

UNIQUENESS THEOREM. *If we assume that the functional* $v \to J(v)$ *is strictly convex, i.e., if*

$$J((1-\theta)u + \theta v) < (1-\theta)J(u) + \theta J(v) \quad \textit{if} \quad 0 < \theta < 1, \quad u \neq v, \tag{1.7}$$

then (*obviously*) *there is at most one u satisfying* (1.6).

Let us give now more "analytic" conditions for (1.6) to hold true. If we assume that the function $v \to J(v)$ is *differentiable*,[4] then (1.6) is *equivalent to*

$$\begin{aligned} &(J'(u), v - u) \geqq 0 \quad \text{for all} \quad v \in U_{ad}, \\ &u \in U_{ad}, \end{aligned} \tag{1.8}$$

---

[1] Actually U will be a (real) *Hilbert space* in most of the applications we shall consider here.

[2] We already use the notation of Control Theory; see Chapter 2.

[3] $\|\cdot\|$ = norm in U. Condition (1.5) is irrelevant if U_{ad} is bounded.

[4] Cf. J. Dieudonné [1], Chap. 8, § 1.

where

$$(J'(u), \phi) = \frac{d}{d\lambda} J(u + \lambda\phi)|_{\lambda=0}.$$

If we assume that J takes the form:

$$J(v) = J_1(v) + J_2(v), \tag{1.9}$$

where J_1 is *differentiable* and J_2 is *continuous* (not necessarily differentiable), and where both functions are convex, then (1.6) is equivalent to

$$(1.10)\quad \begin{aligned} &(J_1'(u), v - u) + J_2(v) - J_1(u) \geqq 0 \quad \text{for all} \quad v \in U_{\text{ad}}, \\ &u \in U_{\text{ad}}. \end{aligned}$$

Extensions: monotone operators and variational inequalities. When $v \to J(v)$ is a convex differentiable function from $U \to \mathbb{R}$, then its derivative J' is a *monotone* operator from $U \to U'$ (dual space of U), i.e.,

$$(J'(u) - J'(v), u - v) \geqq 0 \quad \text{for all} \quad u, v \in U. \tag{1.11}$$

But there are monotone operators from $U \to U'$ which are *not* derivatives of convex functionals (for this kind of question, cf. R. T. Rockafellar [1]); it is then natural to consider a monotone operator $v \to A(v)$ from $U \to U'$, and to consider the problem of finding $u \in U_{\text{ad}}$ such that

$$(A(u), v - u) \geqq 0 \quad \text{for all} \quad v \in U_{\text{ad}}. \tag{1.12}$$

This is a so-called "variational inequality."

Example 1.1. Let us suppose that U is a Hilbert space and that

$$A(v) = Av - f, \tag{1.13}$$

where $A \in L(U; U')$[5] and where f is given in U'.

Then if

$$(Av, v) \geqq \alpha\|v\|^2, \quad \alpha > 0, \quad \text{for all} \quad v \in U_{\text{ad}}, \tag{1.14}$$

there exists a unique $u \in U_{\text{ad}}$ such that

$$(Au - f, v - u) \geqq 0 \quad \text{for all} \quad v \in U_{\text{ad}}. \tag{1.15}$$

For the proof, cf. Lions and Stampacchia [1].

Remark 1.1. If A is *symmetric*, then $Au - f$ is the derivative of

$$J(v) = \tfrac{1}{2}(Av, v) - (f, v);$$

if A is *not* symmetric, then (1.15) does not correspond to a minimization problem. But (1.15) is then a useful tool.

---

[5] $L(X; Y)$ denotes the space of continuous linear mappings from $U \to U'$.

Remark 1.2. The result (1.15) can be considerably extended to cases when A is *not* a linear operator. We refer to Hartman and Stampacchia [1], Brézis [1], J. L. Lions [2].

2. Unilateral problems.

2.1. A model "unilateral" problem in elasticity. Unilateral problems in elasticity theory were introduced by Signorini [1] and studied in Fichera [1] (cf. also Lions and Stampacchia [1]).

We consider here only a very simple model: let Ω be an open set in $\mathbb{R}^n$, Γ its boundary (assumed to be smooth). We look for a function u defined in Ω such that

$$-\Delta u + a_0 u = f \quad \text{in } \Omega, \quad f \in L^2(\Omega), \tag{2.1}$$

where a_0 is greater than 0, subject to the following nonlinear boundary conditions:

$$u \geqq 0 \quad \text{on } \Gamma,$$

$$\frac{\partial u}{\partial n} \geqq 0 \quad \text{on } \Gamma \quad \text{(where } \partial/\partial n = \text{normal derivative } \textit{directed toward the exterior of } \Omega\text{)}, \tag{2.2}$$

$$u\frac{\partial u}{\partial n} = 0 \quad \text{on } \Gamma.$$

We can apply the results of § 1 to show that this problem admits a unique solution.

We introduce

$$U = H^1(\Omega) = \left\{\phi | \phi, \frac{\partial \phi}{\partial x_1}, \cdots, \frac{\partial \phi}{\partial x_n} \in L^2(\Omega)\right\}; \tag{2.3}$$

provided with the norm

$$\|\phi\| = \left(|\phi|^2 + \sum_{i=1}^{n} \left|\frac{\partial \phi}{\partial x_i}\right|\right)^{1/2},$$

where $|\phi| = \left(\int_\Omega \phi^2 \, dx\right)^{1/2}$, $H^1(\Omega)$ is a Hilbert space. This is the *Sobolev space* of order 1 (cf. Sobolev [1]).

For $\phi, \psi \in H^1(\Omega)$, we set

$$a(\phi, \psi) = \sum_{i=1}^{n} \int_\Omega \frac{\partial \phi}{\partial x_i} \frac{\partial \psi}{\partial x_i} dx + \int_\Omega a_0 \phi \psi \, dx. \tag{2.4}$$

We consider next[6]

$$U_{\text{ad}} = \{\phi | \phi \in U, \phi \geqq 0 \text{ on } \Gamma\}. \tag{2.5}$$

[6] We use in this definition the fact that for $\phi \in U$ one can define the trace of ϕ on Γ (cf. for instance Lions and Magenes [1], vol. 1).

One can easily check that U_{ad} is a *closed* convex set in U. Applying (1.15) with $(A\phi, \psi) = a(\phi, \psi)$ for all $\phi, \psi \in U$, we obtain the existence and uniqueness of $u \in U_{\text{ad}}$ such that

$$a(u, \phi - u) \geqq (f, \phi - u) = \int_\Omega f(\phi - u)\, dx \quad \text{for all} \quad \phi \in U_{\text{ad}}. \tag{2.6}$$

It is now a simple exercise to verify that (2.6) is equivalent to (2.1)–(2.2) which proves the existence and uniqueness of u as the solution of (2.1)–(2.2).

Remark 2.1. From the last condition (2.2) it follows that there is a region $\Gamma_0 \subset \Gamma$ where $u = 0$ and that $\partial u/\partial n = 0$ on $\Gamma - \Gamma_0$. But Γ_0 is *not* known a priori and actually finding u is equivalent to finding Γ_0. Therefore one can think of this problem as a *free boundary problem.* Another example is given in § 2.2.

Remark 2.2. Let us consider a subset E of Γ of positive measure, and let us denote by u_E the solution of

$$-\Delta w + q_0 w = f,$$

$$w = 0 \quad \text{on } E, \quad \frac{\partial w}{\partial n} = 0 \quad \text{on } \Gamma - E.$$

Then (cf. Y. Haugazeau [1])

$$u = \sup_E u_E. \tag{2.7}$$

For the proofs of all the statements given up to now, cf. Lions [1].

2.2. A free boundary problem. Let us introduce now

$$H_0^1(\Omega) = \{\phi \mid \phi \in H^1(\Omega),\ \phi = 0 \text{ on } \Gamma\}, \tag{2.8}$$

which is a closed subspace of $H^1(\Omega)$, and let us take $U = H_0^1(\Omega)$ and

$$U_{\text{ad}} = \{\phi \mid \phi \in H_0^1(\Omega),\ \phi \geqq \psi_0 \text{ a.e. in } \Omega\}, \tag{2.9}$$

where ψ_0 is given in $H^1(\Omega)$ and $\psi_0 < 0$ on Γ (in order that U_{ad} not be empty). We take again $a(\phi, \psi)$ given by (2.6). There exists a unique u satisfying (2.6) (with U_{ad} given now by (2.9)). It is a simple matter to check that this problem is equivalent to finding u such that

$$\begin{aligned} &u - \psi_0 \geqq 0 \quad \text{in } \Omega, \\ &-\Delta u + a_0 u - f \geqq 0 \quad \text{in } \Omega, \\ &(u - \psi_0)(-\Delta u + a_0 u - f) = 0 \quad \text{in } \Omega, \\ &u = 0 \quad \text{on } \Gamma. \end{aligned} \tag{2.10}$$

We see that there is in Ω a region (cf. Fig. 1) where $u = \psi_0$ and $u > \psi_0$ in the other part of Ω, with a *free boundary* at the interface.

For the study of the regularity of the solution we refer to H. Brézis and G. Stampacchia [1] and to H. Lewy and G. Stampacchia [1].

For the solution of another free boundary problem arising in hydromechanics we refer to C. Baiocchi [1].

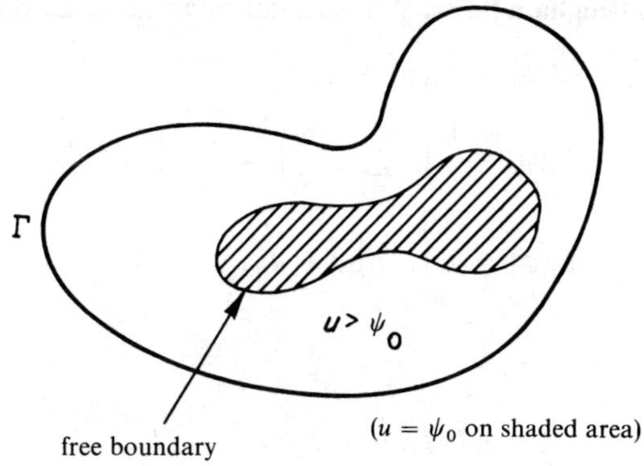

FIG. 1

3. Problems in plasticity. Let us consider, with the notation of § 2,

$$J(\phi) = \frac{1}{2}\int_\Omega \sum_{i=1}^{n} \left(\frac{\partial \phi}{\partial x_i}\right)^2 dx - \int_\Omega f\phi\, dx, \tag{3.1}$$

$$U = H_0^1(\Omega) \tag{3.2}$$

and let us consider

$$U_{\mathrm{ad}} = \{\phi \mid |\mathrm{grad}\, \phi(x)| \leqq 1 \text{ a.e. in } \Omega\}. \tag{3.3}$$

The problem

$$\inf J(v) \quad \text{for all} \quad v \in U_{\mathrm{ad}}, \tag{3.4}$$

arises in plasticity theory (cf. W. Prager and P. G. Hodge [1], Duvaut and Lions [1], Chap. 5, and the bibliographies therein). If we set

$$a(\phi, \psi) = \sum_{i=1}^{n} \int_\Omega \frac{\partial \phi}{\partial x_i}\frac{\partial \psi}{\partial x_i}\, dx, \tag{3.5}$$

then (3.4), which admits a unique solution u, is equivalent to

$$a(u, \phi - u) \geqq (f, \phi - u) \quad \text{for all} \quad \phi \in U_{\mathrm{ad}}. \tag{3.6}$$

We can check that the solution u satisfies $|\mathrm{grad}\, u| = 1$ in the "plastic" region (cf. shaded region in Fig. 2) and satisfies the usual equation $-\Delta u = f$ in the "elastic" region, the two regions being separated by a "free boundary" (the elasto-plastic boundary).

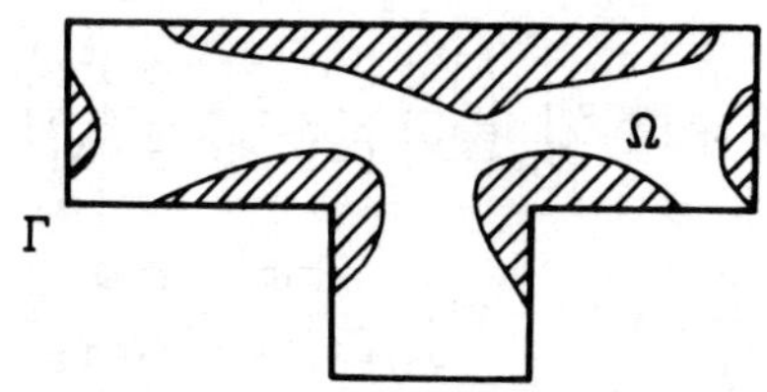

FIG. 2

4. Problems in Bingham flows. We consider now the functional

$$J(\phi) = J_1(\phi) + J_2(\phi),$$

(4.1)
$$J_1(\phi) = \frac{1}{2}\int_\Omega \sum_{i=1}^{n} \left(\frac{\partial \phi}{\partial x_i}\right)^2 dx - \int_\Omega f\phi\, dx,$$

$$J_2(\phi) = g\int_\Omega |\text{grad}\, \phi|\, dx, \quad g > 0,$$

on $U = H_0^1(\Omega)$.

The problem

(4.2)
$$\inf\, [J_1(v) + J_2(v)], \quad v \in U,$$

arises in connection with flows of some non-Newtonian fluids, i.e., the so-called Bingham's fluids. (cf. Duvaut and Lions [1], Chap. 6, and the bibliography therein).

The solution u can be characterized by

(4.3)
$$a(u, \phi - u) + J_2(\phi) - J_2(u) \geqq (f, \phi - u) \quad \text{for all} \quad \phi \in U$$

by application of (1.10).

5. Problems in plate theory. For the motivation (thin plates with friction at the boundary), we refer to Duvaut and Lions [1], Chap. 4.

We introduce the Sobolev space of order 2:

(5.1)
$$H^2(\Omega) = \left\{\phi | \phi, \frac{\partial \phi}{\partial x_i}, \frac{\partial^2 \phi}{\partial x_i \partial x_j} \in L^2(\Omega), i, j = 1, \cdots, n\right\},$$

which is a Hilbert space when provided with the norm

(5.2)
$$\|\phi\| = \left(|\phi|^2 + \sum_{i=1}^{n} \left|\frac{\partial \phi}{\partial x_i}\right|^2 + \sum_{i,j=1}^{n} \left|\frac{\partial^2 \phi}{\partial x_i \partial x_j}\right|^2\right)^{1/2}.$$

We take $\Omega \subset \mathbb{R}^2$ and we introduce

(5.3)
$$U = \{\phi | \phi \in H^2(\Omega), \phi = 0 \text{ on } \Gamma\} = H^2(\Omega) \cap H_0^1(\Omega).$$

For $\phi, \psi \in U$, we set

(5.4)
$$\begin{aligned} a(\phi, \psi) = {} & \sum_{i=1}^{2} \int_\Omega \frac{\partial^2 \phi}{\partial x_i^2} \frac{\partial^2 \psi}{\partial x_i^2}\, dx \\ & + \nu \int_\Omega \left(\frac{\partial^2 \phi}{\partial x_1^2} \frac{\partial^2 \psi}{\partial x_2^2} + \frac{\partial^2 \phi}{\partial x_2^2} \frac{\partial^2 \psi}{\partial x_1^2}\right) dx \\ & + (1 - \nu) \int_\Omega \frac{\partial^2 \phi}{\partial x_1 \partial x_2} \frac{\partial^2 \psi}{\partial x_1 \partial x_2}\, dx, \end{aligned}$$

where $0 < \nu < \frac{1}{2}$.

Green's formula gives, for all $\phi, \psi \in U$:

$$(\Delta^2\phi, \psi) = a(\phi, \psi) + \int_\Gamma (M\phi)\frac{\partial\psi}{\partial n}\,d\Gamma, \tag{5.5}$$

where M is a partial differential operator of second order. We define:

$$\begin{aligned} J_1(\phi) &= \tfrac{1}{2}a(\phi, \phi) - (f, \phi), \\ J_2(\phi) &= k\int_\Gamma \left|\frac{\partial\phi}{\partial n}\right| d\Gamma,\ k > 0, \\ J(\phi) &= J_1(\phi) + J_2(\phi). \end{aligned} \tag{5.6}$$

The problem we consider is:

$$\inf J(v), \quad v \in U. \tag{5.7}$$

This problem admits a unique solution u (cf. Duvaut and Lions [1], Chap. 4) which is characterized by

$$a(u, \phi - u) + J_2(\phi) - J_2(u) \geqq (f, \phi - u) \quad \text{for all} \quad \phi \in U. \tag{5.8}$$

We can check that (5.8) is equivalent to

$$\Delta^2 u = f \quad \text{in } \Omega, \tag{5.9}$$

subject to the boundary conditions

$$u = 0 \quad \text{on } \Gamma \tag{5.10}$$

and

$$|Mu| \leqq k, \quad k\left|\frac{\partial u}{\partial n}\right| - (Mu)\frac{\partial u}{\partial n} = 0. \tag{5.11}$$

The condition (5.11) is equivalent to

$$\begin{aligned} |Mu| < k \quad &\text{implies} \quad \frac{\partial u}{\partial n} = 0, \\ Mu = k \quad &\text{implies} \quad \frac{\partial u}{\partial n} \geqq 0, \\ Mu = -k \quad &\text{implies} \quad \frac{\partial u}{\partial n} \leqq 0. \end{aligned} \tag{5.12}$$

CHAPTER 2

Problems of Optimal Control for Linear Distributed Parameter Systems

1. General considerations. Let Λ be a linear partial differential operator. For the time being, we can think of Λ as a continuous linear operator from Y into F, where Y and F are, say, Hilbert spaces. The space Y contains in its definition some appropriate boundary conditions. We assume that the corresponding boundary value problem is *well-set*, i.e., Λ is an isomorphism from Y onto F.

Let U be the space of controls. We assume that U is a (real) Hilbert space. Let U_{ad} be a closed convex subset of U (the set of *admissible* controls).

Let $B \in L(U; F)$. For every $v \in U$, we consider the (abstract) boundary value problem

$$\Lambda y = f + Bv. \tag{1.1}$$

Equation (1.1) *uniquely defines the state y of the system*:

$$y = y(v) = \Lambda^{-1}(f + Bv). \tag{1.2}$$

Let $C \in L(Y; H)$ (H = Hilbert space) be the "observation" operator; we consider the *cost function*

$$J(v) = \|Cy(v) - z_d\|^2 + (Nv, v)_U, \tag{1.3}$$

where z_d is given in H and $N \in L(U; U)$ is a *symmetric positive definite operator.* The problem is to minimize J on U_{ad}:

$$\inf J(v), \quad v \in U_{\mathrm{ad}}. \tag{1.4}$$

Since $v \to Cy(v)$ is linear + constant from $U \to H$, and $(Nv, v) \geqq \nu\|v\|_U^2$, $\nu > 0$, the function $v \to J(v)$ is continuous convex and even *strictly* convex and $J(v) \to +\infty$ if $\|v\|_U \to \infty$. Consequently, there exists a unique element u in U_{ad} such that

$$J(u) = \inf J(v), \quad v \in U_{\mathrm{ad}}; \tag{1.5}$$

u is the *optimal control.*

It follows directly that J is differentiable; therefore (see Chap. 1, § 1) u is characterized by (1.8) of Chap. 1, which can be written (after dividing by 2):

$$(Cy(u) - z_d, C(y(v) - y(u))) + (Nu, v - u)_U \geqq 0 \quad \text{for all} \quad v \in U_{\mathrm{ad}}. \tag{1.6}$$

Let us introduce the adjoint C^* of C:

$$C^* \in L(H; Y'), \tag{1.7}$$

and the *adjoint state* $p = p(u)$, defined by

$$\Lambda^* p = C^*(Cy(u) - z_d), \tag{1.8}$$

where Λ^* denotes the adjoint operator of Λ.

Then

$$\begin{aligned}(C^*(Cy(u) - z_d), y(v) - y(u)) &= (Cy(u) - z_d, C(y(v) - y(u))) \\ &= (\Lambda^* p, y(v) - y(u)) = (p, \Lambda y(v) - \Lambda y(u)) \\ &= (p, B(v - u)) = (B^* p, v - u)_U\end{aligned}$$

so that (1.6) is equivalent to $(B^*p + Nu, v - u)_U \geqq 0$ for all $v \in U_{\text{ad}}$.

Summing up, we have: the optimal control u, which exists and is unique, is given by the solution of the "optimality system":[1]

$$\begin{aligned}&\Lambda y = f + Bu, \\ &\Lambda^* p = C^*(Cy - z_d), \\ &u \in U_{\text{ad}}, \\ &(B^*p + Nu, v - u)_U \geqq 0 \quad \text{for all} \quad v \in U_{\text{ad}}.\end{aligned} \tag{1.9}$$

In (1.9), y and p are subject to appropriate boundary conditions.

We shall now extend by examples the above (trivial) remarks; the main problem is to analyze more closely the structure and the properties of the corresponding "optimality system".

Remark 1.1. The above presentation was introduced in Lions [3]. The problems correspond to deterministic systems with non-noisy observations. In the case of noisy observations, one can use, following Bensoussan [1], [2] cylindrical probabilities on the Hilbert spaces. For other considerations along these lines, cf. P. Krée [1].

2. Example of an elliptic system. We consider in this section a very simple example, as an easy exercise in applying the principles of § 1. For more general problems for systems governed by elliptic operators we refer to Lions [1], Chap. 2 and to Lions and Magenes [1].

2.1. State equation and optimality system. Let $\Omega \subset \mathbb{R}^n$ be an open set and let Γ be its boundary. We suppose that the state $y = y(v) = y(x; v)$ is given by

$$-\Delta y + y = f \quad \text{in } \Omega, \quad f \in L^2(\Omega), \tag{2.1}$$

$$\frac{\partial y}{\partial n} = v \quad \text{on } \Gamma \tag{2.2}$$

($\partial/\partial n$ denotes the normal derivative directed toward the exterior of Ω).

[1] Where we set $y(u) = y$, $p(u) = p$.

If $v \in U = L^2(\Gamma)$, the system (2.1)–(2.2) uniquely defines $y \in H^1(\Omega)$; the variational form of the problem is

(2.3) $$a(y, \phi) = (f, \phi) + \int_\Omega v\phi \, d\Gamma \quad \text{for all} \quad \phi \in H^1(\Omega),$$

where

(2.4) $$a(\phi, \psi) = \int_\Gamma \left(\sum_{i=1}^n \frac{\partial \phi}{\partial x_i} \frac{\partial \psi}{\partial x_i} + \phi\psi \right) dx.$$

We suppose that the cost function is given by

(2.5) $$J(v) = \int_\Gamma |y(v) - z_d|^2 \, d\Gamma + N \int_\Gamma v^2 \, d\Gamma, \quad N > 0, \quad z_d \in L^2(\Gamma).$$

Then (this is an exercise in applying Green's formula, cf. details in a more general setting in Lions [1], Chap. 2, § 2.4) the *optimality system* is

(2.6) $$\begin{aligned} -\Delta y + y &= f, \\ -\Delta p + p &= 0 \quad \text{in } \Omega, \end{aligned}$$

(2.7) $$\begin{gathered} \frac{\partial y}{\partial n} = u, \quad \frac{\partial p}{\partial n} = y - z_d, \\ u \in U_{\text{ad}}, \quad \int_\Gamma (p + Nu)(v - u) \, d\Gamma \geqq 0 \quad \text{for all} \quad v \in U_{\text{ad}}. \end{gathered}$$

Remark 2.1. Let us emphasize that the optimality system (2.6)–(2.7) being equivalent to the original problem admits a unique solution.

2.2. Example 2.1. No constraints. If $U_{\text{ad}} = U(= L^2(\Gamma))$ (the "no constraints" case), then the last inequalities in (2.7) reduce to

(2.8) $$p + Nu = 0,$$

and eliminating u the optimality system becomes (2.6) subject to boundary conditions:

(2.9) $$\frac{\partial y}{\partial n} + \frac{1}{N} p = 0, \quad \frac{\partial p}{\partial n} = y - z_d.$$

2.3. Example 2.2. Let us suppose now that

(2.10) $$U_{\text{ad}} = \{v | v \in L^2(\Gamma), \quad v \geqq 0 \quad \text{a.e. on } \Gamma\}.$$

Then the inequalities in (2.7) are equivalent to

(2.11) $$u \geqq 0, \quad p + Nu \geqq 0, \quad u(p + Nu) = 0 \quad \text{on } \Gamma.$$

This problem is of the same nature as that of the problems we met in Chapter 1.

Remark 2.2. We can write (2.11) as

(2.11′) $$u = \sup\left(0, -\frac{p}{N}\right) = \frac{1}{N} p^-$$

so that the optimality system becomes (2.6) subject to the nonlinear boundary conditions:

$$\frac{\partial y}{\partial n} = \frac{1}{N}p^-, \quad \frac{\partial p}{\partial n} = y - z_d. \tag{2.12}$$

Remark 2.3. *Regularity result.* We know that the solution $\{y, p\}$ of (2.6), (2.12) belongs to $H^1(\Omega) \times H^1(\Omega)$ but $-\Delta p + p = 0$, $\partial p/\partial n = y - z_d \in L^2(\Gamma)$ implies[2] that $p \in H^{3/2}(\Omega)$ so that $p|_\Gamma \in H^1(\Gamma)$ and $p^-|_\Gamma \in H^1(\Gamma)$ and thus $-\Delta y + y = f$, $\partial y/\partial n \in H^1(\Gamma)$. This implies that $y \in H^{3/2}(\Omega)$. Summing up, we have

$$\{y, p\} \in H^{3/2}(\Omega) \times H^{3/2}(\Omega)$$

and $u \in H^1(\Gamma)$.

3. Systems governed by parabolic equations (I). In this section we consider an example which is a nonstationary analogue of the example studied in § 2. A more general setting is given in § 4.

3.1. State equation and optimality system. Notation is indicated in Fig. 3. We consider the evolution equation

$$\frac{\partial y}{\partial t} - \Delta y = f \quad \text{in} \quad Q = \Omega \times]0, T[, \tag{3.1}$$

subject to the boundary condition

$$\frac{\partial y}{\partial n} = v \quad \text{on} \quad \Sigma = \Gamma \times]0, T[, \tag{3.2}$$

with the initial temperature given:

$$y(x, 0) = y_0(x). \tag{3.3}$$

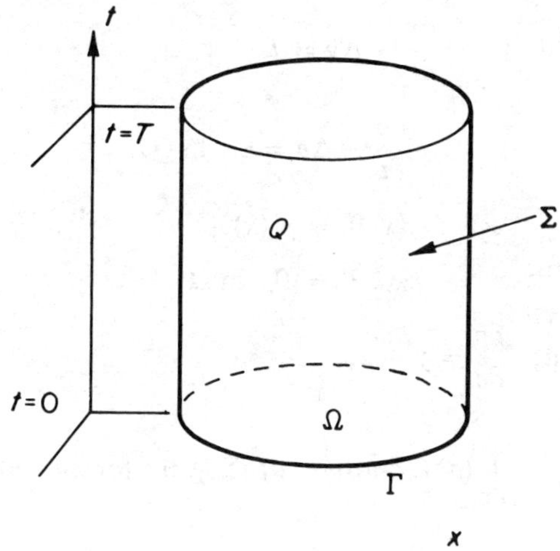

FIG. 3

[2] We use here results of Lions and Magenes [1], vol. 1, Chap. 2.

If $v \in U = L^2(\Sigma)$, then (3.1)–(3.3) uniquely define $y = y(v)$ as an element of $L^2(0, T; H^1(\Omega))$:

(3.4) $$y \in L^2(0, T; H^1(\Omega)),$$

i.e.,

(3.5) $$\int_0^T \|y(t)\|^2_{H^1(\Omega)}\, dt < \infty .$$

The variational formulation of the (classical) problem (3.1)–(3.3) is

(3.6) $$\begin{gathered}\left(\frac{\partial y}{\partial t}, \phi\right) + a(y, \phi) = (f, \phi) + \int_\Gamma v(x, t)\phi(x)\, d\Gamma_x \quad \text{for all} \quad \phi \in H^1(\Omega),\\ y(\cdot\,, 0) = y_0,\end{gathered}$$

where

(3.7) $$\begin{aligned} a(\phi, \psi) &= \int_\Omega \sum_{i=1}^n \frac{\partial \phi}{\partial x_i}\frac{\partial \psi}{\partial x_i}\, dx,\\ (f, \phi) &= \int_\Omega f(x)\phi(x)\, dx. \end{aligned}$$

The cost function is:

(3.8) $$J(v) = \int_\Sigma |y(v) - z_d|^2\, d\Sigma + N\int_\Sigma v^2\, d\Sigma .$$

Then the optimality system is given by:

(3.9) $$\begin{gathered}\frac{\partial y}{\partial t} - \Delta y = f,\\ -\frac{\partial p}{\partial t} - \Delta p = 0 \quad \text{in } Q,\end{gathered}$$

(3.10) $$\begin{gathered} y(x, 0) = y_0(x),\\ p(x, T) = 0 \quad \text{in } \Omega,\end{gathered}$$

(3.11) $$\begin{gathered}\frac{\partial y}{\partial n} = u, \quad \frac{\partial p}{\partial n} = y - z_d,\\ u \in U_{\mathrm{ad}}, \quad \int_\Sigma (p + Nu)(v - u)\, d\Sigma \geqq 0 \quad \text{for all} \quad v \in U_{\mathrm{ad}}.\end{gathered}$$

3.2. Example 3.1. No constraints. If $U_{\mathrm{ad}} = U(= L^2(\Sigma))$, then the last inequality in (3.11) reduces to

(3.12) $$p + Nu = 0 \quad \text{on } \Sigma.$$

If we eliminate u, the optimality system becomes (3.9)–(3.10) *subject to the boundary conditions*:

$$\frac{\partial y}{\partial n} + \frac{1}{N}p = 0, \quad \frac{\partial p}{\partial n} = y - z_d. \tag{3.13}$$

Remark 3.1. The problem (3.9)–(3.10), (3.13) is the extension to distributed parameter systems of the well-known "two-point boundary value problem" met in the case of lumped parameter systems (cf. in particular Athans and Falb [1], Lee and Markus [1]).

Remark 3.2. We shall return to the problem (3.9)–(3.11) in § 6 below.

3.3. Example 3.2. Let us suppose now that

$$U_{\mathrm{ad}} = \{v | v \in L^2(\Sigma), \quad v \geqq 0 \quad \text{a.e. on } \Sigma\}. \tag{3.14}$$

Then the inequalities in (3.11) are equivalent to

$$u \geqq 0, \quad p + Nu \geqq 0, \quad u(p + Nu) = 0 \quad \text{on } \Sigma. \tag{3.15}$$

If we eliminate u we finally obtain (3.9), (3.10) *subject to the nonlinear boundary conditions*:

$$\frac{\partial y}{\partial n} = \frac{1}{N}p^-, \quad \frac{\partial p}{\partial n} = y - z_d \quad \text{on } \Sigma. \tag{3.16}$$

We see (cf. Fig. 4) that we are led to distinguish on Σ two regions: Σ_+ where $p > 0$ and Σ_- where $p \leqq 0$, with a "free boundary" at the interface.

4. Systems governed by parabolic equations (II). We now introduce problems of parabolic type in a more general framework.

4.1. State equation. Let V and H be two (real) Hilbert spaces, $V \subset H$, V dense in H, $V \to H$ continuous. We denote by $\| \cdot \|$ (resp. $| \cdot |$) the norm in V (resp. H) and by

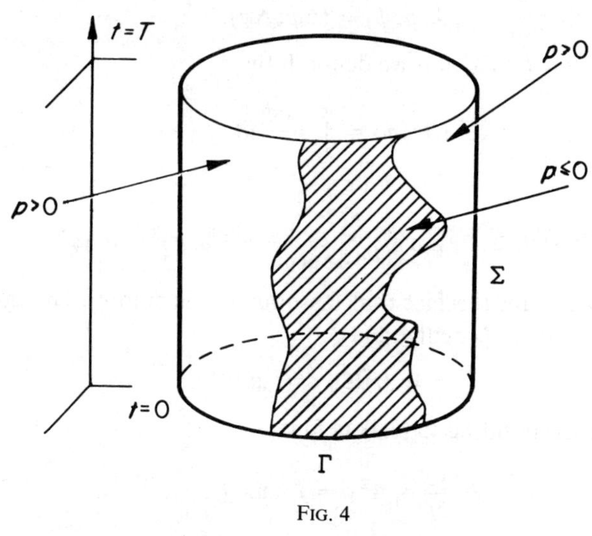

FIG. 4

(,) the scalar product on H. We identify H to its dual so that if V' denotes the dual of V we have

$$V \subset H \subset V'. \tag{4.1}$$

Let $a(\phi, \psi)$ be a continuous bilinear form on V.

We suppose that

$$U = L^2(0, T; \hat{U}), \tag{4.2}$$

where $\hat{U}$ = real Hilbert space, and we consider

$$B \in L(\hat{U}; V'). \tag{4.3}$$

Then, for f given in $L^2(0, T; V')$, y_0 given in H, and v given in U, we consider the *evolution equation*:

$$\begin{aligned} &\left(\frac{\partial y(t)}{\partial t}, \phi\right) + a(y(t), \phi) = (f(t), \phi) + (Bv(t), \phi) \quad \text{for all} \quad \phi \in V, \\ &y(t) \in V, \end{aligned} \tag{4.4}$$

with the initial condition

$$y(0) = y_0, \quad y_0 \text{ given in } H. \tag{4.5}$$

Example 4.1. Let $V = H^1(\Omega)$, $H = L^2(\Omega)$, $\hat{U} = L^2(\Gamma)$; let a be given by (3.7) and B be given by

$$(Bv, \phi) = \int_\Gamma v\phi \, d\Gamma.$$

We then obtain the situation of the examples in § 3.

Example 4.2. We consider $H = L^2(\Omega)$ and

$$V = \{\phi | \phi, \Delta\phi \in L^2(\Omega), \quad \phi = 0 \quad \text{on } \Gamma\}, \tag{4.6}$$

$$a(\phi, \psi) = (\Delta\phi, \Delta\psi). \tag{4.7}$$

We take again $\hat{U} = L^2(\Gamma)$, and we define B by

$$(Bv, \phi) = \int_\Gamma v \frac{\partial \phi}{\partial n} d\Gamma. \tag{4.8}$$

We have

$$|(Bv, \phi)| \leqq \|v\|_{L^2(\Gamma)} \left\| \frac{\partial \phi}{\partial n} \right\|_{L^2(\Gamma)} \leqq c\|v\|_{L^2(\Gamma)} \|\phi\|_{H^2(\Omega)};$$

hence (4.3) follows, using the fact that (assuming Γ is smooth enough) one has (regularity theorem; cf. L. Nirenberg [1]):

$$V = H^2(\Omega) \cap H_0^1(\Omega).$$

The problem corresponding to (4.4)–(4.5) is

$$\frac{\partial y}{\partial t} + \Delta^2 y = f \quad \text{in } Q, \tag{4.9}$$

(4.10) $$y = 0, \quad \Delta y = v \quad \text{on } \Sigma,$$

(4.11) $$y(x, 0) = y_0(x) \quad \text{in } \Omega.$$

If we return to the general theory, we assume that there exist $\lambda \in \mathbb{R}$ and $\alpha > 0$, such that

(4.12) $$a(\phi, \phi) + \lambda|\phi|^2 \geqq \alpha\|\phi\|^2 \quad \text{for all} \quad \phi \in V.$$

Then (cf. Lions [7], Lions and Magenes [1], vol. 1, Chap. 3) the problem (4.4)–(4.5) admits a unique solution $y = y(t, v) = y(v)$ which satisfies

(4.13) $$y \in L^2(0, T; V),$$

(4.14) $$\frac{\partial y}{\partial t} \in L^2(0, T; V').$$

4.2. Optimality system. We consider

(4.15) $$C \in L(V; H), \quad N \in L(\hat{U}; \hat{U}) \text{ symmetric positive definite,}$$

and the cost function

(4.16) $$J(v) = \int_0^T \|Cy(t, v) - z_d(t)\|_H^2 \, dt + \int_0^T (Nv(t), v(t))_{\hat{U}} \, dt,$$

where z_d is given in $L^2(0, T; H)$.

If U_{ad} is a closed convex subset of U, the problem is:

(4.17) $$\inf J(v), \quad v \in U_{\text{ad}},$$

which admits a unique solution u. If we introduce the adjoint state p by

(4.18) $$-\left(\frac{\partial p}{\partial t}, \phi\right) + a^*(p, \phi) = (C^*(C_y - z_d), \phi) \quad \text{for all} \quad \phi \in V,$$

(4.19) $$p(T) = 0,$$

where

(4.20) $$a^*(\phi, \psi) = a(\psi, \phi) \quad \text{for all} \quad \phi, \psi \in V,$$

then the optimality condition becomes

(4.21) $$\int_0^T (p(t), B(v(t) - u(t))) \, dt + \int_0^T (Nu(t), v(t) - u(t))_{\hat{U}} \, dt \geqq 0 \quad \text{for all} \quad v \in U_{\text{ad}}$$

i.e.,

(4.22) $$\int_0^T (B^*p + Nu, v - u)_{\hat{U}} \, dt \geqq 0 \text{ for all} \quad v \in U_{\text{ad}}.$$

Summing up, the optimality system is given by

$$\left(\frac{\partial y}{\partial t}, \phi\right) + a(y, \phi) = (f + Bu, \phi) \quad \text{for all} \quad \phi \in V,$$

(4.23)

$$-\left(\frac{\partial p}{\partial t}, \phi\right) + a^*(p, \phi) = (C^*(Cy - z_d), \phi) \quad \text{for all} \quad \phi \in V,$$

(4.24) $$y(0) = y_0, \quad p(T) = 0,$$

and (4.22).

4.3. An example. We consider the situation of Example 4.1, system (4.9)–(4.11) and the cost function:

(4.25) $$J(v) = \int_\Sigma \left|\frac{\partial}{\partial n} y(v) - z_d\right|^2 d\Sigma + N \int_\Sigma v^2 \, d\Sigma.$$

The adjoint state is given by

(4.26) $$-\frac{\partial p}{\partial t} + \Delta^2 p = 0 \quad \text{in } Q,$$

(4.27) $$p = 0, \quad \Delta p = \frac{\partial y}{\partial n} - z_d \quad \text{on } \Sigma,$$

(4.28) $$p(x, T) = 0 \quad \text{in } \Omega.$$

Then the optimal condition is

(4.29)
$$u \in U_{\mathrm{ad}},$$
$$\int_\Sigma (p + Nu)(v - u) \, d\Sigma \geqq 0,$$

and the optimality system is given by (4.9), (4.10) (where $v = u$), (4.11) and (4.26)–(4.29).

Remark 4.1. For other examples, we refer to Lions [1].

5. Pointwise controls. In the above sections we considered boundary controls and boundary observations. In the present section, we consider controls concentrated on internal points in Ω and distributed observation.

5.1. The state equation. We suppose that the space-dimension n is $\leqq 3$—which is enough for applications! Let $b_1, \cdots, b_m$ be given points in Ω and let us denote by $\delta(x - b_j)$ the Dirac measure at point b_j. The state y of the system is given by

(5.1) $$\frac{\partial y}{\partial t} - \Delta y = f + \sum_{j=1}^{m} v_j(t)\delta(x - b_j) \quad \text{in } Q,$$

(5.2) $$y = 0 \quad \text{on } \Sigma,$$

(5.3) $$y(x, 0) = y_0(x) \quad \text{in } \Omega.$$

In (5.1) f is given in $L^2(Q)$ and $v_j \in L^2(0, T)$. The control is

$$v = \{v_1, \cdots, v_m\} \in U = (L^2(0, T))^m. \tag{5.4}$$

The set of equations (5.1)-(5.3) uniquely defines $y = y(v)$ in $L^2(Q)$.

To check this, let us first proceed formally; let ϕ be given in $L^2(Q)$ and let ψ denote the solution of

$$\begin{gathered} -\frac{\partial \psi}{\partial t} - \Delta\psi = \phi \quad \text{in } Q, \\ \psi = 0 \quad \text{on } \Sigma, \quad \psi(x, T) = 0 \quad \text{in } \Omega. \end{gathered} \tag{5.5}$$

Then we multiply (5.1) by ψ and we integrate over Q. Integrating by parts and taking into account the boundary conditions, we end up with:

$$\int_Q y\phi \, dx\, dt = \int_Q y\psi \, dx\, dt + \int_\Omega y_0(x)\psi(x, 0)\, dx + \sum_{j=1}^{m} \int_0^T v_j(t)\psi(b_j, t)\, dt. \tag{5.6}$$

Let us denote by $L(\psi)$ the right-hand side of (5.6). If we admit for a moment that[3]

$$\phi \to L(\psi)$$

is continuous on $L^2(Q)$, then (5.6) uniquely defines $y \in L^2(Q)$. But the solution ψ of (5.5) satisfies

$$\psi \in L^2(0, T; H^2(\Omega)), \quad \frac{\partial \psi}{\partial t} \in L^2(Q) \tag{5.7}$$

and

$$\|\psi\|_{L^2(0,T;H^2(\Omega))} + \left\| \frac{\partial \psi}{\partial t} \right\|_{L^2(Q)} \leqq c\|\phi\|_{L^2(Q)}. \tag{5.8}$$

But (5.7) implies that, in particular,

$$\psi(\cdot, 0) \in L^2(\Omega). \tag{5.9}$$

Since $n \leqq 3$, $H^2(\Omega) \subset C^0(\bar{\Omega})$ (space of continuous functions on $\bar{\Omega}$) and therefore $\psi(b_j, t) \in L^2(0, T)$ and

$$\int_0^T |\psi(b_j, t)|^2\, dt \leqq c\|\phi\|^2_{L^2(Q)}. \tag{5.10}$$

Thus

$$\begin{aligned} |L(\psi)| &\leqq c\left[\|\psi\|_{L^2(Q)} + \|\psi(\cdot, 0)\|_{L^2(\Omega)} + \sum_{j=1}^{m} \|\psi(b_j, t)\|_{L^2(0,T)} \right] \\ &\leqq c\|\phi\|_{L^2(Q)}, \end{aligned}$$

which proves our assertion.

---

[3] Let us recall that ψ is uniquely defined by (5.5).

Moreover, we can easily check that the linear + constant mapping $v \to y(v)$ is *continuous* from $(L^2(0, T))^m \to L^2(Q)$.

Remark 5.1. The above method for solving (5.1)–(5.3) is the *transposition method*, which is extensively used in solving boundary value problems (cf. Lions and Magenes [1]).

5.2. Cost functions. Optimality system. We suppose that the cost function is given by

$$(5.11)\qquad J(v) = \int_Q |y(x,t;v) - z_d(x,t)|^2\,dx\,dt + \sum_{j=1}^m N_j \int_0^T v_j^2\,dt, \qquad N_j > 0,\quad z_d \text{ given in } L^2(Q).$$

The adjoint state is then given by

$$(5.12)\qquad \begin{aligned} -\frac{\partial p}{\partial t} - \Delta p &= y - z_d \quad \text{in } Q,\\ p &= 0 \quad \text{on } \Sigma,\\ p(x,T) &= 0 \quad \text{on } \Omega. \end{aligned}$$

The optimal system is then given by (5.1) (where $v_j = u_j$, $u = \{u_j\}$ = optimal control), (5.2), (5.3), (5.12) and

$$(5.13)\qquad \sum_{j=1}^m \int_0^T (p(b_j,t) + N_j u_j(t))(v_j(t) - u_j(t))\,dt \geqq 0 \quad \text{for all} \quad v \in U_{\text{ad}}.$$

Example 5.1 (Case without constraints). If we suppose that $U_{\text{ad}} = U$, then (5.13) reduces to

$$(5.14)\qquad p(b_j,t) + N_j u_j(t) = 0, \qquad j = 1, \cdots, m.$$

The optimality system is therefore given by

$$(5.15)\qquad \begin{aligned} &\frac{\partial y}{\partial t} - \Delta y + \sum_{j=1}^m \frac{1}{N_j} p(b_j,t)\delta(x - b_j) = f,\\ &-\frac{\partial p}{\partial t} - \Delta p - y = -z_d,\\ &y = p = 0 \quad \text{on } \Sigma,\\ &y(x,0) = y_0(x), \quad p(x,T) = 0 \quad \text{in } \Omega. \end{aligned}$$

Example 5.2. Let us suppose now that

$$(5.16)\qquad U_{\text{ad}} = \{v \mid v_j(t) \geqq 0 \text{ a.e. in } (0,T)\}.$$

Then (5.13) becomes

$$u_j \geqq 0, \quad p(b_j, t) + N_j u_j(t) \geqq 0, \quad u_j(p(b_j, t) + N_j u_j(t)) = 0, \tag{5.17}$$

i.e.,

$$u_j = \frac{1}{N_j} p(b_j, t)^- . \tag{5.18}$$

The optimality system is the nonlinear boundary value problem:

$$\begin{aligned} &\frac{\partial y}{\partial t} - \Delta y - \sum_{j=1}^{m} \frac{1}{N_j} p(b_j, t)^- \delta(x - b_j) = f, \\ &-\frac{\partial p}{\partial t} - \Delta p - y = -z_d, \\ &y = p = 0 \quad \text{on } \Sigma, \\ &y(x, 0) = y_0(x), \quad p(x, T) = 0 \quad \text{in } \Omega. \end{aligned} \tag{5.19}$$

6. Integro partial differential equations of Riccati type. In the cases *without constraints* the linear boundary value problems met in the preceding sections can be "reduced" to nonlinear boundary value problems of somewhat different character; this is given in detail in Lions [1], Chapters 3 and 4. We explain here the method in the situation of Example 5.1 (the corresponding result being not explicitly given in Lions, loc. cit.).

6.1. The method for uncoupling (5.15). Let us consider in the interval $s < t < T$, the system

$$\begin{aligned} &\frac{\partial \phi}{\partial t} - \Delta \phi + \sum_{j=1}^{m} \frac{1}{N_j} \psi(b_j, t) \delta(x - b_j) = f, \\ &-\frac{\partial \psi}{\partial t} - \Delta \psi - \phi = -z_d, \\ &\phi(x, s) = h(x), \quad h \text{ arbitrarily given}, \\ &\psi(x, T) = 0, \\ &\phi = \psi = 0 \quad \text{on} \quad \Gamma \times]s, T[. \end{aligned} \tag{6.1}$$

This system of equations admits a unique solution since it is the optimality system of a problem identical to (5.11) but with $(0, T)$ replaced by (s, T) and y_0 replaced by h. Consequently, given h, (6.1) uniquely defines $\psi(\cdot, s)$. We can check that

$$h \to \psi(\cdot, s)$$

is a *continuous* linear + constant mapping from $L^2(\Omega) \to L^2(\Omega)$ and $H_0^1(\Omega) \to H_0^1(\Omega)$. Consequently,

$$\psi(\cdot, s) = P(s)h + r(s),$$

$$P(s) \in L(L^2(\Omega); L^2(\Omega)) \cap L(H_0^1(\Omega); H_0^1(\Omega)), \tag{6.2}$$

$$r(s) \in H_0^1(\Omega).$$

But if in (6.1) we choose $h = y(\cdot, s)$, then, of course, $\phi = y$, $\psi = p$ in (s, T) and therefore $\psi(\cdot, s) = p(\cdot, s)$, so that (6.2) gives

$$p(\cdot, s) = P(s)y(\cdot, s) + r(s).$$

Let us write $y(s), \cdots$, instead of $y(\cdot, s), \cdots$; since this equality is valid for every s, we have proved that there exists a family of operators $P(t) \in L(L^2(\Omega); L^2(\Omega)) \cap L(H_0^1(\Omega); H_0^1(\Omega))$ such that

$$(P(t)\phi, \psi) = (\phi, P(t)\psi) \quad \text{for all} \quad \phi, \psi \in L^2(\Omega),^4$$
$$(P(t)\phi, \phi) \geqq 0 \tag{6.3}$$

and

$$p(t) = P(t)y(t) + r(t). \tag{6.4}$$

The problem now is to find equations which permit a direct computation of $P(t)$ and $r(t)$. This is done in § 6.2 below.

Remark 6.1. Using L. Schwartz's kernel theorem (L. Schwartz [1]) we have obtained the existence (and uniqueness) of "kernels" $P(x, \xi, t)$ (which are distributions over $\Omega_x \times \Omega_\xi$) such that

$$P(t)h = \int_\Omega P(x, \xi, t)h(\xi)\, d\xi. \tag{6.5}$$

In what follows, we are going to obtain a nonlinear P.D.E. which characterizes $P(x, \xi, t)$.

6.2. The Riccati P.D.E. We can prove (cf. Lions [1]) that we can take the time derivative of identity (6.4). We obtain, using (6.4) in the second equation (5.15):

$$-\frac{\partial P}{\partial t}y - P\frac{\partial y}{\partial t} - \frac{\partial r}{\partial t} - \Delta(Py + r) - y = -z_d. \tag{6.6}$$

In (6.6) we replace $\partial y/\partial t$ by its value obtained from the first equation in (5.15); we obtain

$$-\frac{\partial P}{\partial t}y - P\left(\Delta y - \sum_{j=1}^{m}\frac{1}{N_j}p(b_j, t)\delta(x - b_j)\right) - Pf$$
$$- \frac{\partial r}{\partial t} - \Delta(Py + r) - y = -z_d. \tag{6.7}$$

---

4 Cf. Lions [1] for a proof of (6.3).

But using (6.4) and (6.5) we have

$$p(b_j, t) = \int_\Omega P(b_j, \xi, t) y(\xi, t)\, d\xi + r(b_j, t) \tag{6.8}$$

and

$$P(\delta(x - b_j)) = \int_\Omega P(x, \xi, t)\delta(\xi - b_j)\, d\xi = P(x, b_j, t). \tag{6.9}$$

Therefore (6.7) gives

$$-\frac{\partial P}{\partial t} y - P\Delta y + \sum_{j=1}^{m} \frac{1}{N_j} p(b_j, t) P(x, b_j, t) - Pf - \frac{\partial r}{\partial t} - \Delta(Py + r) - y = -z_d,$$

and using (6.8), we finally obtain

$$\begin{aligned} &-\frac{\partial P}{\partial t} y - P\Delta y - \Delta P y + \sum_{j=1}^{m} \frac{1}{N_j} P(x, b_j, t) \int_\Omega P(b_j, \xi, t) y(\xi, t)\, d\xi \\ &\quad - y - \frac{\partial r}{\partial t} - \Delta r + \sum_{j=1}^{m} \frac{1}{N_j} P(x, b_j, t) r(b_j, t) - Pf = -z_d. \end{aligned} \tag{6.10}$$

But this is an identity in y; hence,

$$-\frac{\partial P}{\partial t} - (\Delta_x + \Delta_\xi)P + \sum_{j=1}^{m} \frac{1}{N_j} P(x, b_j, t) P(b_j, \xi, t) = \delta(x - \xi), \tag{6.11}$$

$$-\frac{\partial r}{\partial t} - \Delta r + \sum_{j=1}^{m} \frac{1}{N_j} P(x, b_j, t) r(b_j, t) = Pf - z_d \tag{6.12}$$

$$\left(= \int_\Omega P(x, \xi, t) f(\xi, t)\, d\xi - z_d \right).$$

The boundary conditions are:

$$P(x, \xi, t) = 0 \quad \text{if} \quad x \in \Gamma, \xi \in \Omega \quad \text{or} \quad x \in \Omega, \xi \in \Gamma, \tag{6.13}$$

$$r(x, t) = 0 \quad \text{if} \quad x \in \Gamma, \tag{6.14}$$

and since $p(x, T) = 0$ we have:

$$P(x, \xi, T) = 0 \quad \text{on} \quad \Omega \times \Omega, \tag{6.15}$$

$$r(x, T) = 0 \quad \text{on} \quad \Omega. \tag{6.16}$$

One can prove the *uniqueness* of the solution of the above systems of equations. Summing up, there exists a unique kernel $P(x, \xi, t)$ in $\Omega \times \Omega \times]0, T[$, which is characterized as the solution of the nonlinear P.D.E. of parabolic type (6.11) subject to boundary condition (6.13) and with the "initial" condition (6.15);[5]

[5] Since we integrate backward in time. We have also to add (6.3).

then the function $r(x, t)$ is the solution of the (linear) parabolic equation (6.12) subject to boundary condition (6.14) and initial condition (6.16).

The optimal control is finally given by

$$u_j(t) = -\frac{1}{N_j}\int_\Omega P(b_j, \xi, t)y(\xi, t)\,d\xi - \frac{1}{N_j}r(b_j, t). \tag{6.17}$$

Remark 6.2. The Riccati P.D.E. that we have just obtained is the extension to distributed systems of the well-known Riccati O.D.E. met in lumped parameter systems (cf. Kalman and Bucy [1]). For filtering theory in distributed parameter systems, we refer to Bensoussan [1], [2].

Remark 6.3. One can ask whether one can solve *directly* (i.e., without going through control theory) the nonlinear P.D.E. (6.11) subject to (6.14), (6.16). This appears to be an open problem for (6.11); for other types of Riccati P.D.E. met in Lions [1], direct approaches are indeed possible: two different methods have been introduced by G. Da Prato [1]-[3] and by R. Temam [1].

Remark 6.4. For other aspects, one can also consult Bucy and Joseph [1], and H. J. Kushner [1].

7. Cases where the control depends only on t. The reduced Riccati P.D.E.

7.1. The state equation and the cost function. We consider the general situation of § 4.1, with $Bv(t)$ replaced by $v(t)w$, where $v(t)$ is a *scalar* function of t alone and where w is a *fixed* element of V. In other words, the state y is given by the solution of

$$\left(\frac{\partial y}{\partial t}, \phi\right) + a(y(t), \phi) = (f(t), \phi) + v(t)(w, \phi) \quad \text{for all} \quad \phi \in V, \tag{7.1}$$

$$y(0) = y_0. \tag{7.2}$$

We suppose that

$$v \in U_{\mathrm{ad}} = \text{closed convex subset of } L^2(0, T)(= U). \tag{7.3}$$

The cost function is given by (we use the notation of (4.16))

$$J(v) = \int_0^T \|Cy(t, v) - z_d\|_H^2\,dt + N\int_0^T v(t)^2\,dt, \qquad N > 0. \tag{7.4}$$

The adjoint state is given by (4.17), (4.19) and the optimality condition becomes

$$\int_0^T [(p(t), w) + Nu(t)](v(t) - u(t))\,dt \geqq 0 \quad \text{for all} \quad v \in U_{\mathrm{ad}}. \tag{7.5}$$

Finally, the optimality system is given by

$$\text{(7.6)}\qquad \begin{aligned} &\left(\frac{\partial y}{\partial t}, \phi\right) + a(y(t), \phi) = (f, \phi) + u(t)(w, \phi) \quad \text{for all} \quad \phi \in V, \\ &-\left(\frac{\partial p}{\partial t}, \phi\right) + a^*(p, \phi) = (C^*(Cy - z_d), \phi) \quad \text{for all} \quad \phi \in V, \\ &y(0) = y_0, \quad p(T) = 0, \end{aligned}$$

and (7.5).

7.2. The case without constraint. If in (7.5) we take $U_{\mathrm{ad}} = U$, we obtain

$$\text{(7.7)}\qquad (p, w) + Nu = 0,$$

and therefore one can eliminate u from (7.6), (7.7), to obtain:

$$\text{(7.8)}\qquad \begin{aligned} &\left(\frac{\partial y}{\partial t}, \phi\right) + a(y, \phi) + \frac{1}{N}(p, w)(w, \phi) = (f, \phi), \\ &-\left(\frac{\partial p}{\partial t}, \phi\right) + a^*(p, \phi) - (C^*Cy, \phi) = -(C^*z_d, \phi), \\ &y(0) = y_0, \quad p(T) = 0. \end{aligned}$$

We are now going to "uncouple" (7.8) by using the method of § 6 and we shall show that there is a great simplification of the problem when w is an eigenfunction, i.e.,

$$\text{(7.9)}\qquad \begin{aligned} &a(w, \phi) = \lambda(w, \phi) \quad \text{for all} \quad \phi \in V, \\ &w \in V. \end{aligned}$$

The method of § 6 applies; we have the existence of $P(t)$ and $r(t)$ such that

$$\text{(7.10)}\qquad p(t) = P(t)y(t) + r(t).$$

If we introduce A and A^* in $L(V; V')$ by

$$a(\phi, \psi) = (A\phi, \psi), \quad a^*(\phi, \psi) = (A^*\phi, \psi),$$

the equations in (7.8) can be written

$$\text{(7.11)}\qquad \begin{aligned} &\frac{\partial y}{\partial t} + Ay + \frac{1}{N}(p, w)w = f, \\ &-\frac{\partial p}{\partial t} + A^*p - C^*Cy = -C^*z_d. \end{aligned}$$

The identification, after bringing (7.10) into (7.11), leads to

$$\text{(7.12)}\qquad -\frac{\partial P}{\partial t}y + PAy + A^*Py + \frac{1}{N}(Py, w)Pw = C^*Cy$$

and

$$-\frac{\partial r}{\partial t} + A^*r + \frac{1}{N}(r, w)Pw = Pf - C^*z_d, \tag{7.13}$$

with, of course,

$$P(T) = 0, \quad r(T) = 0. \tag{7.14}$$

Since (7.11) is an identity in y, we can take $y(t) = w$ (independent of t).

Let us define:

$$P(t)w = Q(t) \in V. \tag{7.15}$$

Then (7.12) gives, using (7.9) (i.e., $Aw = \lambda w$):

$$-\frac{\partial Q}{\partial t} + (A^* + \lambda)Q + \frac{1}{N}(Q, w)Q = C^*Cw, \tag{7.16}$$

$$Q(t) \in V, \quad (Q(t), w) \geqq 0,^6 \tag{7.17}$$

$$Q(T) = 0. \tag{7.18}$$

The equations (7.16)-(7.18) uniquely define $Q(t)$. Let us check that, if f is given by

$$f(t) = g(t)w, \quad g = \text{scalar function in } L^2(0, T), \tag{7.19}$$

then we can compute the optimal policy by solving (7.16)-(7.18) first, and then by solving:[7]

$$\begin{aligned} -\frac{\partial r}{\partial t} + A^*r + \frac{1}{N}(r, w)Q &= gQ - C^*z_d, \\ r(T) &= 0. \end{aligned} \tag{7.20}$$

Indeed when f is given by (7.19), the equation (7.13) becomes (7.20) where it suffices to know Q (and not P) in order to be able to compute r; then (7.7) gives

$$\begin{aligned} u &= -\frac{1}{N}(Py + r, w) \\ &= -\frac{1}{N}(y, Pw) - \frac{1}{N}(r, w); \end{aligned}$$

hence,

$$u = -\frac{1}{N}(Q, y) - \frac{1}{N}(r, w). \tag{7.21}$$

[6] Since $Q(t) = P(t)w$, we have $(Q(t), w) = (P(t)w, w) \geqq 0$.

[7] And we even see below that a complete solution of (7.20) is not necessary, and if we only want u, then hypothesis (7.19) is not necessary.

Here we can notice that (r, w) can be computed very easily: indeed it follows from (7.13) that

$$-\frac{d}{dt}(r, w) + (r, Aw) + \frac{1}{N}(r, w)(Q, w) = (Pf, w) - (C^*z_d, w) = (Q, f) - (C^*z_d, w);$$

hence we have the ordinary differential equation:

$$\begin{aligned} &-\frac{d}{dt}(r, w) + \left(\lambda + \frac{1}{N}(Q, w)\right)(r, w) = (Q, f) - (C^*z_d, w), \\ &(r, w)|_{t=T} = 0. \end{aligned} \tag{7.22}$$

Let us also notice that (Q, w) can be directly computed by an ordinary Riccati equation; indeed, it follows from (7.16) that

$$\begin{aligned} &-\frac{d}{dt}(Q, w) + 2\lambda(Q, w) + \frac{1}{N}(Q, w)^2 = \|Cw\|_H^2, \\ &(Q, w)|_{t=T} = 0, \quad (Q, w) \geqq 0. \end{aligned} \tag{7.23}$$

Remark 7.1. In the particular case when

$$C^*Cw = \mu w, \quad A^*w = \lambda w, \tag{7.24}$$

then (7.16)–(7.18) is *trivial*; in fact we can look for Q in the form

$$Q(t) = \Phi(t)w, \quad \Phi(t) = \text{scalar function}; \tag{7.25}$$

then (7.16) reduces to the ordinary Riccati equation[8]

$$-\frac{d}{dt}\Phi(t) + 2\lambda\Phi + \frac{1}{N}\Phi^2 = \mu, \quad \Phi(T) = 0. \tag{7.26}$$

7.3. Examples.

Example 7.1. Let the state y be given by the solution of

$$\frac{\partial y}{\partial t} - \Delta y = f + v(t) \quad \text{in} \quad \Omega \times]0, T[, \quad f = f(t), \tag{7.27}$$

$$\frac{\partial y}{\partial n} = 0 \quad \text{on } \Sigma, \tag{7.28}$$

$$y(x, 0) = y_0(x) \quad \text{in } \Sigma. \tag{7.29}$$

We suppose that the cost function is given by

$$J(v) = \int_\Sigma |y(v) - z_d|^2 \, d\Sigma + N \int_0^T v^2 \, dt. \tag{7.30}$$

[8] We suppose that $|w| = 1$.

We can apply the preceding remarks, taking $V = H^1(\Omega)$, $w = 1$, $\lambda = 0$, $H = L^2(\Omega)$, C = trace operator from $H^1(\Omega) \to L^2(\Gamma)$. Then

$$(C^*Cy, \phi) = \int_\Gamma y\phi \, d\Gamma$$

and (7.16) becomes

$$\text{(7.31)} \quad -\left(\frac{\partial Q}{\partial t}, \phi\right) + a(Q, \phi) + \frac{1}{N}\left(\int_\Omega Q \, dx\right)(Q, \phi) = \int_\Gamma \phi \, d\Gamma \quad \text{for all} \quad \phi \in H^1(\Omega)$$

$$\left(\text{where } a(\theta, \phi) = \sum_{i=1}^{n} \int_\Omega \frac{\partial \theta}{\partial x_i} \frac{\partial \phi}{\partial x_i} dx\right),$$

or

$$\text{(7.32)} \quad \begin{aligned} &-\frac{\partial Q}{\partial t} + (-\Delta Q) + \frac{1}{N}\left(\int_\Omega Q \, dx\right) Q = 0, \\ &\frac{\partial Q}{\partial n} = 1, \quad \int_\Omega Q \, dx \geqq 0, \\ &Q(x, T) = 0. \end{aligned}$$

Remark 7.2. We shall give in § 7.5 below a direct solution of (7.32).

Example 7.2. Let the state be given now by the solution of

$$\text{(7.33)} \quad \begin{aligned} &\frac{\partial y}{\partial t} - \Delta y = f + v(t)w \quad \text{in} \quad \Omega \times]0, T[, \\ &y = 0 \quad \text{on } \Sigma, \end{aligned}$$

with the initial condition (7.29), where in (7.33) we have

$$\text{(7.34)} \quad -\Delta w = \lambda w, \quad w = 0 \quad \text{on } \Gamma.$$

We take

$$\text{(7.35)} \quad C = \frac{1}{\text{meas } \Omega} \int_\Omega \phi(x) \, dx = \bar{\phi} \quad (\text{mean value of } \phi).$$

Then (7.16)–(7.18) become

$$\text{(7.36)} \quad \begin{aligned} &-\frac{\partial Q}{\partial t} + (-\Delta + \lambda)Q + \frac{1}{N}\left(\int_\Omega Q(x, t)w(x) \, dx\right) Q = \bar{w}, \\ &Q = 0 \quad \text{on } \Sigma, \quad \int_\Omega Q(x, t)w(x) \, dx \geqq 0, \\ &Q(x, T) = 0. \end{aligned}$$

A direct solution of (7.36) is given in § 7.5 below.

Example 7.3. Let the state be given now by

$$\frac{\partial y}{\partial t} + \Delta^2 y = f + v(t)w, \qquad y = 0, \quad \frac{\partial y}{\partial n} = 0 \quad \text{on } \Sigma \tag{7.37}$$

and the initial condition (7.29), where

$$\Delta^2 w = \lambda w, \quad w, \frac{\partial w}{\partial n} = 0 \quad \text{on } \Gamma. \tag{7.38}$$

We again take C given by (7.35). Then (7.16)–(7.18) become:

$$-\frac{\partial Q}{\partial t} + \Delta^2 Q + \frac{1}{N}\left(\int_\Omega Q(x,t)w(x)\,dx\right)Q = \overline{w},$$
$$Q, \frac{\partial Q}{\partial n} = 0 \quad \text{on } \Sigma, \quad \int_\Omega Q(x,t)w(x)\,dx \geqq 0, \tag{7.39}$$
$$Q(x,T) = 0 \quad \text{in } \Omega.$$

This problem admits a unique solution. We do not know a *direct* proof of this fact.

7.4. Behavior as $T \to \infty$. In the case when

$$a(\phi,\phi) \geqq \alpha\|\phi\|^2 \quad \text{for all} \quad \phi \in V, \quad \alpha > 0, \tag{7.40}$$

then we can prove (as in Lions [1], Chap. 3) that

$$Q(t) \to Q_\infty \quad \text{in } H \text{ weakly as} \quad t \to +\infty, \tag{7.41}$$

where Q_∞ satisfies

$$(A^* + \lambda)Q_\infty + \frac{1}{N}(Q_\infty, w)Q_\infty = C^*Cw, \qquad Q_\infty \in V, \quad (Q_\infty, w) \geqq 0. \tag{7.42}$$

Remark 7.3. Equation (7.42) is trivial if $C^*Cw = \mu w$, $A^*w = \lambda w$. We look for Q_∞ in the form

$$Q_\infty = \xi w, \quad \xi \in \mathbb{R}, \tag{7.43}$$

whence (assuming that $|w| = 1$)

$$2\lambda\xi + \frac{1}{N}\xi^2 = \mu, \qquad \xi \geqq 0, \tag{7.44}$$

which uniquely defines ξ.

We do not know if (7.42) *uniquely* defines Q_∞.

Example 7.4. The result (7.42), applied in the situation of Example 7.3, shows the existence of $Q_\infty \in H_0^2(\Omega)$, solution of

$$(7.45)\qquad \begin{aligned} &(\Delta^2 + \lambda)Q_\infty + \frac{1}{N}\left(\int_\Omega Q_\infty(x)w(x)\,dx\right)Q_\infty = \bar{w},\\ &Q_\infty \in H_0^2(\Omega), \quad \int_\Omega Q_\infty(x)w(x)\,dx \geqq 0. \end{aligned}$$

7.5. Direction solution of some of the nonlinear boundary value problems of § 7.3. In this section we want to prove directly the existence and uniqueness of $Q \in L^2(0, \infty; H_0^1(\Omega))$ such that $\partial Q/\partial t \in L^2(0, \infty; H^{-1}(\Omega))$[9] and the solution of (7.36).[10]

In order slightly to simplify the exposition, we change the time orientation and we solve

$$(7.46)\qquad \begin{aligned} &\frac{\partial Q}{\partial t} + (-\Delta + \lambda)Q + \frac{1}{N}\left(\int_\Omega Qw\,dx\right)Q = \bar{w},\\ &Q = 0 \quad \text{on } \Sigma, \quad \int_\Omega Qw\,dx \geqq 0, \quad Q(x, 0) = 0, \end{aligned}$$

or, in variational form,

$$(7.47)\qquad \begin{aligned} &\left(\frac{\partial Q}{\partial t}, \phi\right) + a(Q, \phi) + \lambda(Q, \phi) + \frac{1}{N}(Q, w)(Q, \phi) = (\bar{w}, \phi),\\ &Q(x, 0) = 0, \quad (Q, w) \geqq 0. \end{aligned}$$

We suppose that $w \geqq 0$.

In a first step, we solve

$$(7.48)\qquad \begin{aligned} &\left(\frac{\partial Q}{\partial t}, \phi\right) + a(Q, \phi) + \lambda(Q, \phi) + \frac{1}{N}\left(\int_\Omega |Q|w\,dx\right)(Q, \phi) = (\bar{w}, \phi),\\ &Q(0) = 0. \end{aligned}$$

This is straightforward: taking in (7.48) $\phi = Q$, we obtain

$$(7.49)\qquad \frac{1}{2}\frac{d}{dt}|Q(t)|^2 + a(Q, Q) + \lambda|Q|^2 + \frac{1}{N}\left(\int_\Omega |Q|w\,dx\right)|Q|^2 = (\bar{w}, Q);$$

hence follow a priori estimates, which suffice to prove the existence by compactness arguments (cf. Lions [2], Chap. 1).

---

9 $H^{-1}(\Omega)$ = dual of $H_0^1(\Omega)$.

10 The same method applies to (7.32).

In a second step, we show that the solution Q of (7.48) is $\geqq 0$, which implies that Q actually solves (7.47); we take in (7.48) $\phi = Q^-$,[11] and we obtain

$$-\frac{1}{2}\frac{d}{dt}|Q^-(t)|^2 - a(Q^-, Q^-) - \lambda|Q^-|^2 - \frac{1}{N}\left(\int_\Omega |Q|w\,dx\right)|Q^-|^2 = (\bar{w}, Q^-),$$

$$Q^-(0) = 0;$$

hence it follows that $Q^- = 0$.

Remark 7.4. The same kind of method can be applied to the *stationary* analogue of (7.46).

Remark 7.5. It is clear that the preceding method does not use the fact that $-\Delta w = \lambda w$; what we proved is that the equation

$$\begin{aligned}&\frac{\partial Q}{\partial t} + (-\Delta + \lambda)Q + \left(\int_\Omega Qm\,dx\right)Q = g,\\ &m \in L^2(\Omega), \quad m \geqq 0,\\ &g \in L^2(0, T; H^{-1}(\Omega)), \quad g \geqq 0,\\ &Q = 0 \quad \text{on } \Sigma, \quad Q(x, 0) = 0,\end{aligned} \tag{7.50}$$

admits a unique solution Q which is $\geqq 0$.

It would be interesting to know if one can make similar modifications in (7.39).

7.6. Various remarks.

Remark 7.6. All the considerations made in the preceding sections extend to the case of systems governed by first order hyperbolic operators. The corresponding result proves a conjecture of Kopel, Shih and Conghanowa [1].[12]

Remark 7.7. The same kinds of methods apply to filtering theory. We refer to Bensoussan and Lions [1].

8. Systems governed by hyperbolic or Petrowski operators. We give in § 8.1 an example of a system governed by a hyperbolic operator. For other examples, and for systems governed by Petrowski operators, we refer to Lions [1], Chap. 4 and to Russell [1]–[3] and to the bibliographies therein.

In § 8.2 we give an extension to the remarks of § 7, which leads to apparently new types of nonlinear P.D.E.'s.

---

[11] This method does *not* work in the case of Example 7.3.

[12] At least when, in the notation of these authors, A = diagonal matrix (which is the case in the examples worked out by these authors).

8.1. Example of a system governed by a hyperbolic operator. Let the state y of the system be given by

(8.1) $$\frac{\partial^2 y}{\partial t^2} - \Delta y = f + v \quad \text{in } Q,$$

(8.2) $$\frac{\partial y}{\partial n} = 0 \quad \text{on } \Sigma,$$

(8.3) $$y(x,0) = y_0(x), \quad \frac{\partial y}{\partial t}(x,0) = y_1(x), \quad x \in \Omega.$$

In (8.1) we suppose that $f \in L^2(Q)$ and that $v \in U_{\text{ad}}$, where

(8.4) $$U_{\text{ad}} = \text{closed convex set of } U, \quad U = L^2(Q).$$

We suppose that we observe y on Σ; the cost function is given by

(8.5) $$J(v) = \int_\Sigma |y(v) - z_d|^2 \, d\Sigma + N \int_Q v^2 \, dx \, dt, \quad N > 0.$$

Let u be the optimal control, which exists and is unique; it is characterized by

(8.6) $$\int_\Sigma (y(u) - z_d)(y(v) - y(u)) \, d\Sigma + N \int_Q u(v - u) \, dx \, dt \geqq 0 \quad \text{for all} \quad v \in U_{\text{ad}}.$$

Let us introduce the adjoint state p as the solution of [13]

(8.7) $$\frac{\partial^2 p}{\partial t^2} - \Delta p = 0 \quad \text{in } Q,$$

(8.8) $$\frac{\partial p}{\partial n} = y(u) - z_d \quad \text{on } \Sigma,$$

(8.9) $$p(x,T) = 0, \quad \frac{\partial p}{\partial t}(x,T) = 0 \quad \text{in } \Omega.$$

Then, using Green's formula, we can check that (8.6) is equivalent to

(8.10) $$\int_Q (p + Nu)(v - u) \, dx \, dt \geqq 0 \quad \text{for all} \quad v \in U_{\text{ad}}, \qquad u \in U_{\text{ad}}.$$

The optimality system is given by (8.1) (where $v = u$), (8.2), (8.3) and (8.7)–(8.10).

Example 8.1. Let us suppose that

(8.11) $$U_{\text{ad}} = \{v \mid v \geqq 0 \quad \text{a.e.} \quad \text{in } Q\}.$$

---

[13] For a precise definition of p, one can, for instance, use Lions [1], Chap. 4, § 7.2.

Then (8.10) is equivalent to

$$u = \frac{1}{N} p^- \tag{8.12}$$

so that the optimality system becomes

$$\begin{aligned} &\frac{\partial^2 y}{\partial t^2} - \Delta y - \frac{1}{N} p^- = f, \\ &\frac{\partial^2 p}{\partial t^2} - \Delta p = 0 \quad \text{in } Q, \\ &\frac{\partial y}{\partial n} = 0, \quad \frac{\partial p}{\partial n} = y - z_d \quad \text{on } \Sigma, \\ &y(x,0) = y_0(x), \quad \frac{\partial y}{\partial t}(x,0) = y_1(x), \quad p(x,T) = 0, \quad \frac{\partial p}{\partial t}(x,T) = 0 \quad \text{in } \Omega. \end{aligned} \tag{8.13}$$

8.2. Uncoupling and new types of nonlinear P.D.E.

Notation. We consider the Hilbert spaces V, H as in § 4.1, and we introduce

$$\mathscr{V} = V \times V \subset \hbar = V \times H. \tag{8.14}$$

We suppose in this section that we have *complex* Hilbert spaces. Let $a(\phi, \psi)$ be a continuous sesquilinear form on V. We assume that

$$\begin{aligned} &a(\phi, \psi) = \overline{a(\psi, \phi)}, \\ &a(\phi, \phi) \geqq \alpha \|\phi\|^2, \quad \alpha > 0. \end{aligned} \tag{8.15}$$

We can therefore provide V with the scalar product $a(\phi, \psi)$ and we provide $\hbar$ with the scalar product

$$[\boldsymbol{\phi}, \boldsymbol{\psi}] = a(\phi^0, \psi^0) + (\phi^1, \psi^1), \tag{8.16}$$

where $\boldsymbol{\phi} = \{\phi^0, \phi^1\}, \cdots$.

We identify $\hbar$ with its dual, so that

$$\mathscr{V} \subset \hbar \subset \mathscr{V}' \tag{8.17}$$

and we denote by $[\,,\,]$ the scalar product between $\mathscr{V}$ and $\mathscr{V}'$. We now introduce the operator $\mathscr{A}$ given by

$$\mathscr{A}\boldsymbol{\phi} = \{-\phi^1, A\phi^0\} \tag{8.18}$$

(where $(A\phi, \psi) = a(\phi, \psi)$ for all $\phi, \psi \in V$) which gives

$$\mathscr{A} \in \mathscr{L}(\mathscr{V}; \mathscr{V}'). \tag{8.19}$$

We observe that if $\mathbf{f} = \{0, f\}$ the equation

$$\frac{\partial \mathbf{z}}{\partial t} + \mathscr{A}\mathbf{z} = \mathbf{f}, \quad \mathbf{z}(0) = \mathbf{z}_0 \tag{8.20}$$

is equivalent to

(8.21) $$\frac{\partial^2 z^0}{\partial t^2} + Az^0 = f, \quad z^0(0), \quad \frac{\partial}{\partial t} z^0(0) \quad \text{given}.$$

The equation (8.20) admits a unique solution in $L^\infty(0, T; \mathscr{h})$ if $\mathbf{f} \in L^2(0, T; \mathscr{h})$. (Cf. Lions [7], Lions and Magenes [1], Chap. 3.)

The state of the system. Let us consider $\mathbf{w} \in \mathscr{V}$ such that

(8.22) $$\mathscr{A}\mathbf{w} = i\mu\mathbf{w}, \quad \mu \in \mathbb{R};$$

this amounts to

(8.23) $$-w^1 = i\mu w^0, \quad Aw^0 = i\mu w^1,$$

i.e.,

(8.24) $$Aw^0 = \mu^2 w^0.$$

We take

(8.25) $$U = L^2(0, T)$$

and we suppose that the state $\mathbf{y}$ is given by

(8.26) $$\frac{\partial \mathbf{y}}{\partial t} + \mathscr{A}\mathbf{y} = v(t)\mathbf{w},$$

(8.27) $$\mathbf{y}(0) = \mathbf{y}_0 \quad \text{given in } \mathscr{h}.$$

Cost function. Optimality system.

(8.28) $$J(v) = \int_0^T \|C\mathbf{y}\|_{\mathscr{h}}^2 \, dt + N \int_0^T |v|^2 \, dt, \quad N > 0.$$

We consider the "no constraint function" problem, i.e.,

(8.29) $$\inf J(v), \quad v \in L^2(0, T).$$

Let u denote the unique solution of (8.29). If we introduce the adjoint state, given by

(8.30) $$-\frac{\partial \mathbf{p}}{\partial t} + \mathscr{A}^*\mathbf{p} = C^*C\mathbf{y},$$

(8.31) $$\mathbf{p}(T) = 0,$$

then u is characterized by

(8.32) $$[\mathbf{p}, \mathbf{w}] + Nu = 0.$$

We remark that

(8.33) $$\mathscr{A}^* = -\mathscr{A};$$

indeed

$$[\mathscr{A}^*\boldsymbol{\phi}, \boldsymbol{\psi}] = [\boldsymbol{\phi}, \mathscr{A}\boldsymbol{\psi}] = a(\phi^0, -\phi^1) + (\phi^1, A\phi^0) = -[\mathscr{A}\boldsymbol{\phi}, \boldsymbol{\psi}].$$

Eliminating u, the optimality system is given by

$$(8.34)\qquad \begin{aligned} &\frac{\partial \mathbf{y}}{\partial t} + \mathscr{A}\mathbf{y} + \frac{1}{N}[\mathbf{p}, \mathbf{w}]\mathbf{w} = 0,\\ &-\frac{\partial \mathbf{p}}{\partial t} + \mathscr{A}^*\mathbf{p} - C^*C\mathbf{y} = 0,\\ &\mathbf{y}(0) = \mathbf{y}_0, \quad \mathbf{p}(T) = 0. \end{aligned}$$

Uncoupling of (8.34). Using the method of § 6.1, we obtain the identity

$$(8.35)\qquad \begin{aligned} &\mathbf{p}(t) = \mathscr{P}(t)\mathbf{y}(t),\\ &\mathscr{P}(t) \in \mathscr{L}(\mathscr{V}; \mathscr{V}) \cap \mathscr{L}(\mathscr{h}; \mathscr{h}). \end{aligned}$$

Using (8.35) in (8.34), we obtain:

$$-\left(\frac{\partial \mathscr{P}}{\partial t}\right)\mathbf{y} - \mathscr{P}\frac{\partial \mathbf{y}}{\partial t} + \mathscr{A}^*\mathscr{P}\mathbf{y} - C^*C\mathbf{y} = 0$$

and using the first equation (8.34) we obtain

$$(8.36)\qquad -\left(\frac{\partial}{\partial t}\mathscr{P}\right)\mathbf{y} + \mathscr{P}\mathscr{A}\mathbf{y} + \frac{1}{N}[\mathscr{P}\mathbf{y}, \mathbf{w}]\mathscr{P}\mathbf{w} + \mathscr{A}^*\mathscr{P}\mathbf{y} - C^*C\mathbf{y} = 0.$$

But (8.36) is an identity in $\mathbf{y}$. Therefore we can take $\mathbf{y}(t) = \mathbf{w}$ (independent of t).

Using now the fact that $\mathbf{w}$ satisfies (8.22), after setting

$$(8.37)\qquad \mathscr{P}(t)\mathbf{w} = \mathbf{Q}(t) \in \mathscr{V}$$

we obtain

$$(8.38)\qquad -\frac{\partial \mathbf{Q}}{\partial t} + \mathscr{A}^*\mathbf{Q} + i\mu\mathbf{Q} + \frac{1}{N}[\mathbf{Q}, \mathbf{w}]\mathbf{Q} = C^*C\mathbf{w}.$$

To (8.38) we should of course add (since $\mathbf{p}(T) = 0$)

$$(8.39)\qquad \mathbf{Q}(T) = 0.$$

This shows the existence of $\mathbf{Q} \in L^\infty(0, T; \mathscr{h})$ satisfying (8.38)–(8.39). The uniqueness is easily proved.

If we set $\mathbf{Q} = \{Q^0, Q^1\}$, we have $[\mathbf{Q}, \mathbf{w}] = (\mu^2 Q^0 + i\mu Q^1, w^0)$ so that we have finally proved the existence of uniqueness of $\mathbf{Q} = \{Q^0, Q^1\} \in L^\infty(0, T; V \times H)$ such that

$$(8.40)\qquad -\frac{\partial \mathbf{Q}}{\partial t} - \mathscr{A}\mathbf{Q} + i\mu\mathbf{Q} + \frac{1}{N}(\mu^2 Q^0 + i\mu Q^1, w^0)\mathbf{Q} = C^*C\mathbf{w}$$

with condition (8.39).

Example 8.2. Let us take $V = H_0^1(\Omega)$, $H = L^2(\Omega)$, $A = -\Delta$, $\mathbf{w} = \{w^0, -i\mu w^0\}$,

$$-\Delta w^0 = \mu^2 w^0, \quad w^0 \in H_0^1(\Omega), \tag{8.41}$$

and let C be given by (7.35).

Then (8.40) can be written in the following form:

$$\begin{aligned} -\frac{\partial Q^0}{\partial t} + i\mu Q^0 + Q^1 + \frac{1}{N}\left(\int_\Omega (\mu^2 Q^0 + i\mu Q^1) w^0(x)\,dx\right) Q^0 &= \mathbf{w}^0, \\ -\frac{\partial Q^1}{\partial t} + i\mu Q^1 + \Delta Q^0 + \frac{1}{N}\left(\int_\Omega (\mu^2 Q^0 + i\mu Q^1) w^0(x)\,dx\right) Q^1 &= -i\mu \mathbf{w}^0, \end{aligned} \tag{8.42}$$

with

$$Q^0 = 0 \quad \text{on } \Sigma, \tag{8.43}$$

and

$$Q^0(x, T) = 0, \quad Q^1(x, T) = 0. \tag{8.44}$$

Example 8.3. If we take $V = H_0^2(\Omega)$, $H = L^2(\Omega)$, $A = \Delta^2$, w^0 such that

$$\Delta^2 w^0 = \mu^2 w^0, \quad w^0 \in H_0^2(\Omega), \tag{8.45}$$

and if C is again given by (7.35), then (8.40) can be written

$$\begin{aligned} -\frac{\partial Q^0}{\partial t} + i\mu Q^0 + Q^1 + \frac{1}{N}\left(\int_\Omega (\mu^2 Q^0 + i\mu Q^1) w^0\,dx\right) Q^0 &= \bar{w}^0, \\ -\frac{\partial Q^1}{\partial t} + i\mu Q^1 - \Delta^2 Q^0 + \frac{1}{N}\left(\int_\Omega (\mu^2 Q^0 + i\mu Q^1) w^0\,dx\right) Q^1 &= -i\mu \bar{w}^0, \end{aligned} \tag{8.46}$$

with

$$Q^0 = 0, \quad \frac{\partial Q^0}{\partial n} = 0 \quad \text{on } \Sigma, \tag{8.47}$$

and (8.44).

Remark 8.1. A very large number of examples along these lines can be given. The corresponding nonlinear P.D.E.'s seem to be of new type; a direct solution of these systems of nonlinear P.D.E.'s does not seem to be known.

Remark 8.2. For other aspects of the problems considered in this chapter, such as controllability, time optimal, duality, etc. we refer to Lions [1]. For problems with delays, we refer to J. K. Aggarwal [1], Delfour and Mitter [1], Banks, Jambs and Latina [1], Kushner and Barnes [1], P. J. Reeve [1], A. Bensoussan [3], E. Pardoux [1] and the bibliographies therein.

9. Complement. We can extend a little further the considerations of § 7.[14] We can obtain new systems of nonlinear P.D.E.'s.

[14] We could also extend, along similar lines, the considerations of § 8.

We consider, with the notations of § 7, the equation

$$\frac{\partial y}{\partial t} + Ay = \sum_{i=1}^{q} v_i(t) w_i, \tag{9.1}$$

where $y \in L^2(0, T; V)$, with

$$y(0) = y_0, \tag{9.2}$$

and where

$$v = \{v_i\} \in (L^2(0, T))^q, \tag{9.3}$$

and where the w_i's are eigenfunctions of A:

$$Aw_i = \lambda_i w_i, \qquad 1 \leqq i \leqq q; \quad w_i \in V. \tag{9.4}$$

The cost function $J(v)$ is chosen to be:

$$J(v) = \int_0^T |C^*Cy(v)|^2 \, dt + \sum_{i=1}^{q} N_i \int_0^T v_i^2 \, dt, \quad N_i > 0. \tag{9.5}$$

We consider the case *without constraints*. Let u be the optimal control. If we introduce p by

$$-\frac{\partial p}{\partial t} + A^*p = C^*Cy \tag{9.6}$$

(where $y = y(u)$) and

$$p(T) = 0, \tag{9.7}$$

then u is characterized by

$$\sum_{i=1}^{q} \int_0^T ((p, w_i) + N_i u_i) v_i \, dt = 0 \quad \text{for all} \quad v_i,$$

i.e.,

$$(p, w_i) + N_i u_i = 0, \qquad 1 \leqq i \leqq q. \tag{9.8}$$

By elimination of u_i, we obtain

$$\begin{gathered} \frac{\partial y}{\partial t} + Ay + \sum_{j=1}^{q} \frac{1}{N_j}(p, w_j) w_j = 0, \\ -\frac{\partial p}{\partial t} + A^*p - C^*Cy = 0, \\ y(0) = y_0, \quad p(T) = 0. \end{gathered} \tag{9.9}$$

The uncoupling leads to

$$p = Py$$

and the identification gives:

$$(9.10)\qquad -\left(\frac{\partial P}{\partial t}\right)y + PAy + \sum_{j=1}^{q} \frac{1}{N_j}(Pv, w_j)Pw_j + A^*Py = C^*Cy.$$

If we set

$$(9.11)\qquad Q_j = Pw_j$$

and if in (9.10) we take $y = w_i$, we obtain

$$(9.12)\qquad -\frac{\partial Q_i}{\partial t} + (\lambda_i + A^*)Q_i + \sum_{j=1}^{q} \frac{1}{N_j}(Q_i, w_j)Q_j = C^*Cw_i, \qquad 1 \leqq i \leqq q,$$

with the condition

$$(9.13)\qquad Q_i(T) = 0, \qquad 1 \leqq i \leqq q.$$

This nonlinear system of P.D.E. admits a unique solution $Q_1, \cdots, Q_n$, satisfying

$$(9.14)\qquad Q_i \in L^2(0, T; V).$$

CHAPTER 3

Nonlinear Systems

1. Examples in biomathematics. A large number of examples of optimal control problems for *nonlinear* distributed parameter systems arises in biomathematics (cf. J. P. Kernevez [1], J. P. Kernevez and Thomas [1]). We study here one of these examples.

1.1. State equations. The state y of the system denotes a concentration in a chemical reaction, which takes place in a membrane of width 1 (after normalization the space variable x spans $(0, 1)$); if $i(x, t)$ denotes the concentration of a product which is "slowing down" the reaction, the model can be precisely described by the system of nonlinear P.D.E.'s:

$$\frac{\partial y}{\partial t} - \frac{\partial^2 y}{\partial x^2} + F(y, i) = 0, \qquad 0 < x < 1, \quad 0 < t < T, \tag{1.1}$$

$$\frac{\partial i}{\partial t} - \frac{\partial^2 i}{\partial x^2} = 0, \tag{1.2}$$

where

$$F(y, i) = \sigma \frac{y}{1 + y + \lambda i}, \qquad \sigma, \lambda > 0. \tag{1.3}$$

The boundary conditions are

$$y(0, t) = \alpha(t), \quad y(1, t) = \beta(t), \tag{1.4}$$

where α and β are given positive functions, and

$$i(0, t) = i(1, t) = v(t); \tag{1.5}$$

$v(t)$ is the control variable; in the model (enzymes reactions) one can dispose of v, subject to the constraints

$$0 \leqq v(t) \leqq M, \tag{1.6}$$

by adding water or some very cheap chemical product.

Initial concentrations are zero:

$$y(x, 0) = 0, \quad i(x, 0) = 0. \tag{1.7}$$

Remark 1. *We shall show below that* (1.1)–(1.7) *uniquely define y and i.*

Of course (1.2), (1.5) and the second condition (1.7) uniquely define $i = i(v)$, the mapping $v \to i(v)$ being linear; replacing i by $i(v)$ in $F(y, i)$, the problem reduces to that of solving (1.1), (1.6) and the first condition (1.7).

1.2. Cost functions. We observe $y(\frac{1}{2}, t)$; the cost function is

$$J(v) = \int_0^T |y(\tfrac{1}{2}, t; v) - z_d(t)|^2 \, dt. \tag{1.8}$$

If we define U_{ad} by (1.6), the problem is then:

$$\inf J(v), \quad v \in U_{\text{ad}}. \tag{1.9}$$

In what follows we shall show (§ 1.6) the existence of an optimal control and we shall derive necessary conditions for optimality.

Remark 1.2. The functional $v \to J(v)$ has no reason to have a unique minimum, but a complete study of this point is an open problem.

1.3. Solution of the nonlinear state equations. Let us set

$$g(y) = \sigma \frac{y}{1 + \lambda i(v) + y}, \quad G(y) = \sigma \frac{y}{1 + \lambda i(v) + |y|}. \tag{1.10}$$

Using Remark 1.1, we have to show the existence and uniqueness of y satisfying:

$$y \in L^2(0, T; H^1(\Omega)), \quad \frac{\partial y}{\partial t} \in L^2(0, T; H^{-1}(\Omega))\,[1] \tag{1.11}$$

and

$$\begin{aligned} &\frac{\partial y}{\partial t} - \frac{\partial^2 y}{\partial x^2} + g(y) = 0, \\ &y(0, t) = \alpha(t), \quad y(1, t) = \beta(t), \\ &y(x, 0) = 0. \end{aligned} \tag{1.12}$$

The uniqueness is straightforward. For the existence, let us introduce a function ϕ such that

$$\begin{aligned} &\phi \in L^2(0, T; H^1(\Omega)), \quad \frac{\partial \phi}{\partial t} \in L^2(0, T; H^{-1}(\Omega)), \\ &\phi(0, t) = \alpha(t), \quad \phi(1, t) = \beta(t).\,[2] \end{aligned} \tag{1.13}$$

Then if we set

$$y - \phi = z, \tag{1.14}$$

---

[1] Where $\Omega =]0, 1[$.

[2] We assume that such a function ϕ exists.

we have to solve

$$\frac{\partial z}{\partial t} - \frac{\partial^2 z}{\partial x^2} + g(z + \phi) = -\left(\frac{\partial \phi}{\partial t} - \frac{\partial^2 \phi}{\partial x^2}\right), \quad z(x,0) = -\phi(x,0) \tag{1.15}$$

with

$$z \in L^2(0, T; H_0^1(\Omega)). \tag{1.16}$$

For solving (1.15)–(1.16) we proceed in two steps as in Chap. 2, § 7.5. We consider first the equation

$$\frac{\partial \zeta}{\partial t} - \frac{\partial^2 \zeta}{\partial x^2} + G(\zeta + \phi) = -\left(\frac{\partial \phi}{\partial t} - \frac{\partial^2 \phi}{\partial x^2}\right), \tag{1.17}$$

the other conditions being unchanged. We can prove the existence (and uniqueness) of ζ by standard compactness *or* monotone arguments (cf. Lions [2][3]). We shall therefore obtain the result if we show that

$$\zeta + \phi \geqq 0 \tag{1.18}$$

(since then $G(\zeta + \phi) = g(\zeta + \phi)$ and we can take $z = \zeta$). If we set $\zeta + \phi = m$, we have

$$\frac{\partial m}{\partial t} - \frac{\partial^2 m}{\partial x^2} + G(m) = 0; \tag{1.19}$$

hence, multiplying by m^- and using the fact that

$$m^-(0, t) = m^-(1, t) = 0,^4$$

we obtain

$$-\frac{1}{2}\frac{d}{dt}\int_\Omega (m^-)^2\,dx - \int_\Omega \left(\frac{\partial m^-}{\partial x}\right)^2 dx - \sigma \int_\Omega \frac{(m^-)^2}{1 + \lambda i(v) + |m|}\,dx = 0.$$

Hence $m^- = 0$.[5]

1.4. Existence of an optimal control. Necessary conditions. Let us prove first that when v spans U_{ad} (defined by (1.6)), we have that

(1.20) $i(v)$ remains in a relatively compact subset of $L^2(Q)$ $(Q = \Omega \times]0, T[)$.

Indeed, if ϕ is given in $L^2(Q)$, let us define ψ by

$$-\frac{\partial \psi}{\partial t} - \frac{\partial^2 \psi}{\partial x^2} = \phi, \quad \psi(x, T) = 0, \quad \psi(0, t) = \psi(1, t) = 0; \tag{1.21}$$

[3] Notice that $\zeta \to G(\zeta + \phi)$ is a monotone operator from, say, $L^2(\Omega)$, into itself.

[4] Since $\alpha(t)$, $\beta(t)$ are $\geqq 0$.

[5] Using also the fact that $m^-(x, 0) = 0$.

then (cf. Chap. 2, § 5.1) i is defined by

$$\int_Q i\phi \, dx \, dt = -\int_0^T \left[\frac{\partial\psi}{\partial x}(1,t) - \frac{\partial\psi}{\partial x}(0,t)\right] v(t) \, dt. \tag{1.22}$$

Then it is enough to prove that if ϕ remains in a bounded subset of $L^2(Q)$, then $(\partial\psi/\partial x)(a,t)$, $a = 0, 1$, remains in a relatively compact subset of $L^2(0,T)$. But (1.21) shows that ψ (resp. $\partial\psi/\partial t$) remains in a bounded subset of $L^2(0,T;H^2(\Omega))$ (resp. $L^2(Q)$) so that

(1.23)
$$\frac{\partial\psi}{\partial x} \text{ remains in a bounded set of } L^2(0,T;H^1(\Omega)),$$
$$\frac{\partial}{\partial t}(\partial\psi/\partial x) \text{ remains in a bounded set of } L^2(0,T;H^{-1}(\Omega)).$$

But (1.23) implies the desired result; hence (1.20) follows.

On the other hand, it follows directly that, when v spans U_{ad}, we have that

(1.24) $y(v)$ (resp. $\partial y(v)/\partial t$) remains in a bounded set of $L^2(0,T;H^1(\Omega))$ (resp. $L^2(0,T;H^{-1}(\Omega))$).

Since $H^1(\Omega) \subset C^0([0,1])$ = continuous function in $[0,1]$, we have $y(\frac{1}{2},t;v) \in L^2(0,T)$, so that (1.8) makes sense. Let v_n be a minimizing sequence. Using the fact that U_{ad} is, in particular, a closed *bounded* convex set in $L^2(0,T)$ and using (1.20), (1.24), we can extract a subsequence, still denoted by v_n, such that

(1.25)
$$v_n \to u \quad \text{in } L^2(0,T) \text{ weakly},$$
$$i(v_n) = i_n \to i \quad \text{in } L^2(Q) \text{ strongly and a.e.},$$
$$y(v_n) = y_n \to y \quad \text{in } L^2(0,T;H^1(\Omega)) \text{ weakly},$$
$$\frac{\partial y_n}{\partial t} \to \frac{\partial y}{\partial t} \quad \text{in } L^2(0,T;H^{-1}(\Omega)) \text{ weakly}.$$

But (1.25) implies that

$$\sigma\frac{y_n}{1+\lambda i_n + y_n} \to \sigma\frac{y}{1+\lambda i + y} \quad \text{in } L^2(Q) \text{ strongly}$$

and since $i = i(u)$, we can easily check that $y = y(u)$. The third property in (1.25) shows that

$$y_n(\tfrac{1}{2},t) \to y(\tfrac{1}{2},t) \quad \text{in } L^2(0,T) \text{ weakly}$$

so that

$$\liminf J(v_n) \geqq J(u),$$

which proves that

$$u \in U_{ad}, \quad J(u) = \inf J(v), \quad v \in U_{ad}. \tag{1.26}$$

Necessary conditions. We can check without difficulty that the functions

$$\xi \to y(v + \xi\phi), \quad \xi \to i(v+\xi\phi), \quad (v, \phi \in U_{ad})$$

are differentiable from $\mathbb{R}_\xi \to L^2(Q)$ (for instance, cf. Kernevez [1], Brauner and Penel [1]). Let us set

$$\frac{d}{d\xi} y(v+\xi\phi)\bigg|_{\xi=0} = \hat{y}, \quad \frac{d}{d\xi} i(v+\xi\phi)\bigg|_{\xi=0} = \hat{\imath}. \tag{1.27}$$

It is not difficult to check that $\hat{y}$, $\hat{\imath}$ are given by the solution of:

$$\begin{aligned} &\frac{\partial \hat{y}}{\partial t} - \frac{\partial^2 \hat{y}}{\partial x^2} + \frac{\partial F}{\partial y}(y,i)\hat{y} + \frac{\partial F}{\partial i}(y,i)\hat{\imath} = 0, \\ &\frac{\partial \hat{\imath}}{\partial t} - \frac{\partial^2 \hat{\imath}}{\partial x^2} = 0, \end{aligned} \tag{1.28}$$

$$\begin{aligned} &\hat{y}(0,t) = \hat{y}(1,t) = 0, \\ &\hat{\imath}(0,t) = \hat{\imath}(1,t) = \phi(t), \end{aligned} \tag{1.29}$$

$$\hat{y}(x,0) = 0, \quad \hat{\imath}(x,0) = 0. \tag{1.30}$$

We have then

$$\frac{1}{2}(J'(u),\phi) = \int_0^T (y(\tfrac{1}{2},t;u) - z_d(t))\hat{y}(\tfrac{1}{2},t)\,dt, \tag{1.31}$$

where $\hat{y}$ is given by (1.28)–(1.30), where $y = y(u)$, $i = i(u)$.

We now introduce the adjoint state $\{p, q\}$ by the solution of

$$\begin{aligned} &-\frac{\partial p}{\partial t} - \frac{\partial^2 p}{\partial x^2} + \frac{\partial F}{\partial y}(y,i)p = (y(\tfrac{1}{2},t;u) - z_d(t))\delta(x-\tfrac{1}{2}), \\ &-\frac{\partial q}{\partial t} - \frac{\partial^2 q}{\partial x^2} + \frac{\partial F}{\partial i}(y,i)p = 0, \end{aligned} \tag{1.32}$$

$$p(0,t) = p(1,t) = 0, \quad q(0,t) = q(1,t) = 0, \tag{1.33}$$

$$p(x,T) = q(x,T) = 0. \tag{1.34}$$

Let us multiply the right-hand sides of (1.32) respectively by $\hat{y}$ and $\hat{\imath}$ and integrate over Q; we obtain

$$\begin{aligned} \frac{1}{2}(J'(u),\phi) &= \int_0^T \left(\frac{\partial q}{\partial x}(0,t) - \frac{\partial q}{\partial x}(1,t)\right)\phi(t)\,dt \\ &\quad + \int_Q p\left(\frac{\partial \hat{y}}{\partial t} - \frac{\partial^2 \hat{y}}{\partial x^2} + \frac{\partial F}{\partial y}(y,i)\hat{y} + \frac{\partial F}{\partial i}(y,i)\hat{\imath}\right) dx\,dt \\ &\quad + \int_Q q\left(\frac{\partial \hat{\imath}}{\partial t} - \frac{\partial^2 \hat{\imath}}{\partial x^2}\right) dx\,dt \\ &= \int_0^T \left(\frac{\partial q}{\partial x}(0,t) - \frac{\partial q}{\partial x}(1,t)\right)\phi(t)\,dt \qquad \text{(using (1.28)).} \end{aligned}$$

Conclusion. The optimality system[6] is given by (1.1)–(1.4), (1.5), where $v = u$, (1.32)–(1.34) and

$$\int_0^T \left(\frac{\partial q}{\partial x}(0,t) - \frac{\partial q}{\partial x}(1,t)\right)(v-u)\,dt \geqq 0 \quad \text{for all} \quad v \in U_{\text{ad}}, \tag{1.35}$$

$$u \in U_{\text{ad}}.$$

Remark 1.3. The preceding method of proof is rather general. Other nonlinear systems have been considered in Lions [4] (where the existence of time optimal for nonlinear systems is also proved) and further extensions have been given by F. Bidaut [1].

2. Convexity results.

2.1. An inequality. Let $\lambda \to \phi(\lambda)$ be a continuous function from $\mathbb{R} \to \mathbb{R}$ which satisfies:

$$\phi \text{ is concave increasing}, \quad \lambda\phi(\lambda) \geqq 0. \tag{2.1}$$

Then, given f and $v \in L^2(Q)$, there exists a unique function $y = y(x,t;v) = y(v)$ which is the solution of

$$\frac{\partial y}{\partial t} - \Delta y + \phi(y) = f + v \quad \text{in } Q, \tag{2.2}$$

$$y \in L^2(0,T;H_0^1(\Omega)) \cap L^\infty(0,T;L^2(\Omega)),$$
$$\int_0^T \int_\Omega y\phi(y)\,dx\,dt < \infty, \tag{2.3}$$

$$y(x,0) = y_0(x), \quad y_0 \text{ given in } L^2(\Omega). \tag{2.4}$$

We want to prove that, under the assumption (2.1), we have

$$y((1-\theta)v_1 + \theta v_2) \leqq (1-\theta)y(v_1) + \theta y(v_2) \quad \text{a.e. in } Q, \qquad 0 \leqq \theta \leqq 1, \quad v_1, v_2 \in L^2(Q). \tag{2.5}$$

Proof. Let us set $y(v) = y$, $y(v_i) = y_i$, $i = 1, 2$, and

$$Y = y - (1-\theta)y_1 - \theta y_2. \tag{2.6}$$

We have

$$\frac{\partial Y}{\partial t} - \Delta Y + \phi(y) - (1-\theta)\phi(y_1) - \theta\phi(y_2) = 0, \tag{2.7}$$

$$Y = 0 \quad \text{on } \Sigma,$$
$$Y(x,0) = 0 \quad \text{in } \Omega. \tag{2.8}$$

[6] This system admits a solution, and probably "several" solutions.

We want to show that $Y^+ = 0$. Let us multiply (2.7) by Y^+ and integrate over Ω; we obtain:

$$\frac{1}{2}\frac{\partial}{\partial t}\int_\Omega (Y^+)^2\,dx + \int_\Omega \sum_{i=1}^n \left(\frac{\partial Y^+}{\partial x_i}\right)^2 dx + X = 0, \tag{2.9}$$

where

$$X = \int_\Omega (\phi(y) - (1-\theta)\phi(y_1) - \theta\phi(y_2))Y^+\,dx. \tag{2.10}$$

In (2.10) we integrate over the set of x such that (t being fixed) $Y \geqq 0$, i.e., $y \geqq (1-\theta)y_1 + \theta y_2$; but since ϕ is increasing it follows that $\phi(y) \geqq \phi((1-\theta)y_1 + \theta y_2)$ and since ϕ is concave it follows that $\phi(y) \geqq (1-\theta)\phi(y_1) + \theta\phi(y_2)$; hence $X \geqq 0$. But then (2.9) implies $Y^+ = 0$.

2.2. A nonlinear system with a convex cost function. We now consider the system whose state is given by (2.2)–(2.3) and we consider the cost function

$$J(v) = 2\int_Q y(x,t;v)\,dx\,dt + N\int_Q v^2\,dx\,dt, \quad N > 0. \tag{2.11}$$

If U_{ad} = closed convex set of U, $U = L^2(Q)$, we consider the problem

$$\inf J(v), \quad v \in U_{\mathrm{ad}}. \tag{2.12}$$

By virtue of (2.5), the function $v \to \int_Q y(x,t;v)\,dx\,dt$ is convex and therefore $J(v)$ is *strictly convex*.

It follows easily from (2.2)–(2.3) that

$$\|y(v)\|_{L^2(Q)} \leqq C\|v\|_{L^2(Q)}$$

so that

$$J(v) \geqq N\|v\|^2_{L^2(Q)} - C\|v\|_{L^2(Q)}. \tag{2.13}$$

Moreover if $v_n \to v$ in U, then $y(v_n) \to y(v)$ in $L^2(Q)$ so that J is *continuous*. Therefore, there exists a unique optimal control u.

If we assume now that

$$\lambda \to \phi(\lambda) \quad \text{is} \quad C^1, \tag{2.14}$$

then one can write down the optimality system which characterizes u. If we set, as in § 1,

$$\left.\frac{d}{d\xi}y(v + \xi w)\right|_{\xi=0} = \hat{y},$$

we have:

$$\frac{\partial \hat{y}}{\partial t} - \Delta \hat{y} + \phi'(y)\hat{y} = w \quad \text{in } Q, \tag{2.15}$$

$$\begin{aligned} \hat{y} &= 0 \quad \text{on } \Sigma, \\ \hat{y}(x,0) &= 0 \quad \text{in } \Omega. \end{aligned} \tag{2.16}$$

Let us set $y = y(u)$ and let us introduce the adjoint state p as the solution of

$$-\frac{\partial p}{\partial t} - \Delta p + \phi'(y)p = 1 \quad \text{in } Q, \tag{2.17}$$

$$\begin{aligned} p &= 0 \quad \text{on } \Sigma, \\ p(x,T) &= 0 \quad \text{in } \Omega. \end{aligned} \tag{2.18}$$

We have

$$\frac{1}{2}\frac{d}{d\xi}J(u + \xi w)\bigg|_{\xi=0} = \int_Q \hat{y}\,dx\,dt + N\int_Q uw\,dx\,dt; \tag{2.19}$$

multiplying (2.17) by $\hat{y}$ we obtain, after integrating by parts,

$$\int_Q \hat{y}\,dx\,dt = \int_Q pw\,dx\,dt.$$

Therefore, the optimality system is given by

$$\begin{aligned} &\frac{\partial y}{\partial t} - \Delta y + \phi(y) = f + u, \\ &-\frac{\partial p}{\partial t} - \Delta p + \phi'(y)p = 1 \quad \text{in } Q, \end{aligned} \tag{2.20}$$

$$y = 0, \quad p = 0 \quad \text{on } \Sigma, \tag{2.21}$$

$$y(x,0) = y_0(x), \quad p(x,T) = 0 \quad \text{in } \Omega, \tag{2.22}$$

and

$$\begin{aligned} &\int_Q (p + Nu)(v - u)\,dx\,dt \geqq 0 \quad \text{for all} \quad v \in U_{ad}, \\ &u \in U_{ad}. \end{aligned} \tag{2.23}$$

This system admits a unique solution.

3. Systems governed by partial differential inequalities.[7] We have introduced in Chapter 1 some problems of mechanics which reduced to "variational inequalities"

[7] This section was written after discussions with G. Duvaut. Cf. also Duvaut and Lions [1].

or P. D. inequalities. We consider here the control of a system governed by a P.D. inequality of *stationary type.*[8]

3.1. Problem of optimal load. We consider the problem of Chap. 1, § 5. The state $y = y(v)$ is given by the solution of

$$\Delta^2 y = f + v \quad \text{in } \Omega, \tag{3.1}$$

$$y = 0 \quad \text{on } \Gamma, \tag{3.2}$$

$$|\mathcal{M} y| \leqq k, \quad k\left|\frac{\partial y}{\partial n}\right| - (\mathcal{M} y)\frac{\partial y}{\partial n} = 0 \quad \text{on } \Gamma.^{9} \tag{3.3}$$

We suppose that the total load is given:

$$\int_Q v(x)\,dx = M, \tag{3.4}$$

and that v is subject to

$$0 \leqq v(x) \leqq v_0 \quad \text{a.e. in } \Omega. \tag{3.5}$$

Let us call U_{ad} the set of $v \in L^2(\Omega)$ satisfying (3.4)–(3.5). Here $y(x)$ denotes the deflection of the state at point x and we wish to minimize over U_{ad} the functional:

$$J(v) = \int_\Gamma b\left(\frac{\partial y}{\partial n}(v)\right) d\Gamma, \tag{3.6}$$

where $\lambda \to b(\lambda)$ is a given continuous increasing function. Let us prove the existence of an optimal control.

We recall that $y(v)$ is characterized by (cf. Chap. 1, § 5; we use the notation of this section):

$$\begin{gathered} a(y(v), \phi - y(v)) + J_2(\phi) - J_2(y(v)) \geqq (f + v, \phi - y(v)) \\ \text{for all} \quad \phi \in V = H^2(\Omega) \cap H_0^1(\Omega), \quad y(v) \in V. \end{gathered} \tag{3.7}$$

If v spans a bounded set in $L^2(\Omega)$, it follows from (3.7) that $y(v)$ spans a bounded set in $H^2(\Omega)$. Then

(3.8) $\dfrac{\partial y(v)}{\partial n}$ spans a bounded set in $H^{1/2}(\Gamma)$ and hence a relatively compact set in $L^2(\Gamma)$.[10]

---

[8] One meets in biochemistry the problem of control of a system governed by a P.D. inequality of evolution. We refer to Kernevez [2].

[9] These conditions express friction laws at the boundary.

[10] Cf. Lions and Magenes [1], vol. 1, Chap. 1.

Let now v_j be a minimizing sequence. Let us set $y(v_j) = y_j$. We can extract a subsequence, denoted by v_j, such that

$$v_j \to u \quad \text{in } L^2(\Omega) \text{ weakly}, \quad u \in U_{\text{ad}}, \tag{3.9}$$

$$y_j \to y \quad \text{in } H^2(\Omega) \text{ weakly}, \tag{3.10}$$

$$\frac{\partial y_j}{\partial n} \to \frac{\partial y}{\partial n} \quad \text{in } L^2(\Gamma) \text{ strongly and a.e. on } \Gamma. \tag{3.11}$$

Using (3.7), we have

$$a(y_j, \phi) + J_2(\phi) - J_2(y_j) - (f + v_j, \phi - y_j) \geqq a(y_j, y_j). \tag{3.12}$$

Using (3.9), (3.10), (3.11) (and the fact that $y_j \to y$ in $L^2(\Omega)$ strongly) we obtain from (3.12),

$$a(y, \phi) + J_2(\phi) - J_2(y) - (f + u, \phi - y) \geqq \liminf a(y_j, y_j) \geqq a(y, y);$$

hence $y = y(u)$.

We deduce from this: $\liminf J(v_j) \geqq J(u)$, so that u is an optimal control.

Remark 3.1. The mapping $v \to y(v)$ is *not* differentiable. The question of writing useful necessary conditions for optimality is open.

3.2. A variant of the inequality of § 2.1. The following result is already given in Lions [8].

We suppose that the state of the system is given by

$$\begin{gathered} -\Delta y + y = f \quad \text{in } \Omega, \\ y \geqq 0, \quad \frac{\partial y}{\partial n} - v \geqq 0, \quad y\left(\frac{\partial y}{\partial n} - v\right) = 0 \quad \text{on } \Gamma, \quad v \in L^2(\Gamma). \end{gathered} \tag{3.13}$$

We introduce:

$$\begin{gathered} K = \{\phi | \phi \in H^1(\Omega), \phi \geqq 0 \text{ on } \Gamma\}, \\ a(\phi, \psi) = \int_\Omega \left(\phi\psi + \sum_{i=1}^n \frac{\partial \phi}{\partial x_i}\frac{\partial \psi}{\partial x_i}\right) dx. \end{gathered} \tag{3.14}$$

Then (cf. Chap. 1, § 2.1) (3.13) is equivalent to

$$\begin{gathered} y = y(v) \in K, \\ a(y(v), \phi - y(v)) \geqq (f, \phi - y(v)) + \int_\Gamma v(\phi - y(v))\, d\Gamma \quad \text{for all} \quad \phi \in K. \end{gathered} \tag{3.15}$$

We have

$$y((1-\theta)v_1 + \theta v_2) \leqq (1-\theta)y(v_1) + \theta y(v_2), \qquad 0 \leqq \theta \leqq 1, \quad v_i \in L^2(\Gamma). \tag{3.16}$$

Let us set $y(v) = y$, $y(v_i) = y_i$, $Y = y - (1 - \theta)y_1 - \theta y_2$. We choose

$$\phi = y - Y^+$$

in (3.15)[11] and $\phi = y_1 + (1 - \theta)Y^+$ (resp. $\phi = y_2 + \theta Y^+$) in the inequality analogous to (3.15) with v replaced by v_1 (resp. v_2) and we add up. We obtain $-a(Y, Y^+) \geqq 0$; hence $Y^+ = 0$ and thus (3.16) follows.

Remark 3.2. An application of (3.16) similar to the one given in § 2.2 can be given but now the mapping $v \to y(v)$ is *not* differentiable and the problem of writing necessary conditions is open.

4. Controls in the coefficients of the highest order derivatives.

4.1. Statement of the problem. Problems similar to those considered here appear in K. A. Luré [1]. Let us define

$$U_{\text{ad}} = \{v | 0 < \mu \leqq v \leqq M \text{ a.e. in } \Omega\} \tag{4.1}$$

and let the state $y = y(v)$ of the system be given by the solution of

$$\begin{gathered} -\sum_{i,j=1}^{n} \frac{\partial}{\partial x_i}\left(v a_{ij} \frac{\partial y}{\partial x_j}\right) = f \quad \text{in } \Omega, \\ y = 0 \quad \text{on } \Gamma, \end{gathered} \tag{4.2}$$

where the a_{ij}'s are given in $L^\infty(\Omega)$ and satisfy the ellipticity condition

$$\sum_{i,j=1}^{n} a_{ij}(x)\xi_i\xi_j \geqq \alpha \sum_{i=1}^{n} \xi_i^2 \quad \text{a.e. in } \Omega, \qquad \alpha > 0. \tag{4.3}$$

If we consider the functional

$$J(v) = \|y(v) - z_d\|_{L^2(\Omega)}, \tag{4.4}$$

it is not known if there exists an optimal control; we know of the existence of *relaxed* optimal controls; this has been investigated by L. Cesari [1], in the spirit of the work of L. C. Young [1], E. J. McShane [1], R. V. Gamkrelidze [1], J. Warga [1]. (For other approaches to relaxed optimal control of distributed parameter systems, cf. I. Ekeland [1], Ioffe and Tikomirov [1].)

But L. Tartar [1] has observed that another type of information can be obtained if we consider instead of (4.4):

$$J(v) = \|y(v) - z_d\|_{H^1(\Omega)}; \; z_d \in H^1(\Omega). \tag{4.5}$$

The result is then: there exists a set $Z \subset H^1(\Omega)$, dense in $H^1(\Omega)$, such that for all $z_d \in Z$, the problem

$$\inf J(v), \quad v \in U_{\text{ad}}, \quad J \text{ given by (4.5)}, \tag{4.6}$$

admits a solution.

---

[11] On Γ, $y_i \geqq 0$ so that $y \geqq Y$; hence $y^+ = y \geqq Y^+$, and thus $\phi \geqq 0$ on Γ and $\phi \in K$.

4.2. Proof of the existence of an optimal control for a dense set of z_d. The proof is actually a straightforward application of a result of Edelstein [1]: *if S is a nonempty closed set in a uniformly convex Banach space X, $\inf_{s \in S} \|s - z_d\|$ admits a solution in S for z_d in a dense subset of X.*

We apply this result with $X = H^1(\Omega)$ and

$$S = \{y(v) | v \in U_{\mathrm{ad}}\}. \tag{4.7}$$

The only thing we have to check is that S defined by (4.7) *is closed in* $H^1(\Omega)$ (we do not know if it is closed in $L^2(\Omega)$). But this is straightforward: suppose that $y_n = y(v_n) \to y$ in $H^1(\Omega)$; we can extract a subsequence $v_\mu \to v$ in $L^\infty(\Omega)$ weak-star, and since $\partial y_\mu/\partial x_i \to \partial y/\partial x_i$ in $L^2(\Omega)$ strongly, we have

$$-\sum_{i,j=1}^{n} \frac{\partial}{\partial x_i}\left(v_\mu a_{ij}\frac{\partial y_\mu}{\partial x_j}\right) \to -\sum_{i,j=1}^{n} \frac{\partial}{\partial x_i}\left(v a_{ij}\frac{\partial y}{\partial x_j}\right) \quad \text{in } H^{-1}(\Omega)$$

so that $y = y(v)$; hence the result follows.

Remark 4.1. The main point in Edelstein's result is that S is *not* (necessarily) convex.

Remark 4.2. For some cases where one can prove the analogue of Pontryagin's maximum principle (cf. Pontryagin, Boltyanskii, Gamkrelidze and Mischenko [1]), we refer to W. Fleming [1].

CHAPTER 4

Optimization of Geometrical Elements

1. Optimal position of Dirac measures.

1.1. Setting of the problem. We consider the situation of Chap. 2, § 5.1; the state y is given by

$$\frac{\partial y}{\partial t} - \Delta y = f + \sum_{i=1}^{m} v_j(t)\delta(x - b_j) \quad \text{in } Q, \tag{1.1}$$

$$y = 0 \quad \text{on } \Sigma, \tag{1.2}$$

$$y(x, 0) = y_0(x), \tag{1.3}$$

$\Omega \subset \mathbb{R}^n, n \leqq 3$. We assume that

$$v = \{v_1, \cdots, v_m\} \in U = (L^2(0, T))^m, \tag{1.4}$$

and we suppose also that we have the possibility of choosing b_j, subject to

$$\begin{gathered} b_j \in E, \quad 1 \leqq j \leqq n, \\ E = \text{bounded closed set} \subset \Omega. \end{gathered} \tag{1.5}$$

If we set

$$b = \{b_1, \cdots, b_m\} \in E^m, \tag{1.6}$$

we can write

$$y = y(x, t; v, b) = y(v, b). \tag{1.7}$$

In this framework, the cost function corresponding to (5.11), Chap. 2, can now be written:

$$J(v, b) = \int_Q |y(x, t; v, b) - z_d(x, t)|^2 \, dx \, dt + \sum_{j=1}^{m} N_j \int_0^T |v_j(t)|^2 \, dt. \tag{1.8}$$

The problem is now:

$$\inf J(v, b), \quad v \in U_{\text{ad}}, \quad b \in E^m, \tag{1.9}$$

where U_{ad} is a closed convex set in U.

1.2. An existence result. We are going to prove that there exist $u \in U_{\text{ad}}, b_0 \in E^m$ such that

$$J(u, b_0) \leqq J(v, b) \quad \text{for all} \quad v \in U_{\text{ad}}, \quad \text{for all} \quad b \in E^m. \tag{1.10}$$

Let indeed $v^\alpha = \{v_j^\alpha\}$, $b^\alpha = \{b_j^\alpha\}$ be a minimizing sequence. Since

$$J(v, b) \geqq (\inf_j N_j) \sum_{j=1}^{m} \int_0^T |v_j(t)|^2 \, dt,$$

v^α remains in a bounded set of U_{ad} and we can extract a sequence, still denoted by v^α, b^α such that

$$v^\alpha \to u \quad \text{in } (L^2(0, T))^m \text{ weakly}, \quad u \in U_{\text{ad}}, \tag{1.11}$$

$$b^\alpha \to b_0 \quad \text{in } E^m. \tag{1.12}$$

But then $v_j^\alpha(t)\delta(x - b_j^\alpha) \to u_j(t)\delta(x - b_{0j})$, for instance in the sense of distributions in Q (cf. L. Schwartz [2]). Let us denote by y^α the state corresponding to v^α, b^α. It follows from (5.6), Chap. 2, that y^α remains in a bounded set of $L^2(Q)$, so that we can extract a sequence, still denoted by y^α, such that $y^\alpha \to y$ in $L^2(Q)$ weakly; the equality

$$\int_Q y^\alpha \phi \, dx \, dt = \int_Q f\psi \, dx \, dt + \int_\Omega y_0(x)\psi(x, 0) \, dx + \sum_{j=1}^{m} \int_0^T v_j^\alpha(t)\psi(b_j^\alpha, t) \, dt \tag{1.13}$$

(with the notation of (5.6), Chap. 2) gives at the limit,

$$\int_Q y\phi \, dx \, dt = \int_Q f\psi \, dx \, dt + \int_\Omega y_0(x)\psi(x, 0) \, dx + \sum_{j=1}^{m} \int_0^T u_j(t)\psi(b_{0j}, t) \, dt$$

so that $y = y(u, b_0)$.

Then $\liminf J(v^\alpha, b^\alpha) \geqq J(u, b_0)$; hence the result follows.

2. A priori feedback laws.

2.1. A special feedback law. Let us consider the following problem: the state is given by

$$\frac{\partial y}{\partial t} - \Delta y = f + v \quad \text{in } Q, \tag{2.1}$$

$$y = 0 \quad \text{on } \Sigma, \tag{2.2}$$

$$y(x, 0) = y_0(x) \quad \text{on } \Omega, \tag{2.3}$$

and the cost function is given by

$$\begin{aligned} J(v) &= \int_Q |y(x, t; v) - z_d|^2 \, dx \, dt + \int_Q N(x, t)v(x, t)^2 \, dx \, dt, \\ N(x, t) &\geqq N > 0. \end{aligned} \tag{2.4}$$

The problem is to find

$$\inf J(v), \quad v \in U_{\text{ad}}, \quad U_{\text{ad}} = \text{closed convex set of} \quad U, U = L^2(Q). \tag{2.5}$$

If u denotes the optimal control and if we can compute an approximation of u, it will generally be extremely difficult to implement it. A possible simplification of this point is to consider a priori that v is of a given form, namely,

$$v(x, t) = \sum_{j=1}^{m} v_j(t)\mathcal{M}_j y, \tag{2.6}$$

where $\mathcal{M}_j$ is a suitable operator acting on y.

Thus, we are led to the following problem: let y be the solution[1] of

$$\frac{\partial y}{\partial t} - \Delta y - \sum_{j=1}^{m} v_j(t)\mathcal{M}_j y = f \tag{2.7}$$

subject to (2.2)–(2.3) and let the cost function J be given by

$$J(v) = \int_Q |y(v) - z_d|^2 \, dx \, dt + \sum_{j=1}^{m} N_j \int_0^T |v_j(t)|^2 \, dt, \quad N_j > 0. \tag{2.8}$$

The new problem is to find

$$\inf J(v), \quad v \in U_{\text{ad}}, \quad U_{\text{ad}} = \text{closed convex set of} \quad (L^2(0, T))^m. \tag{2.9}$$

We give a general existence result in § 2.2 and we return to (2.7) in § 2.3.

2.2. A general existence result. We use the notation of Chap. 2, § 4.1.

We consider a family of operators $\mathcal{M}_i$, $1 \leqq i \leqq m$, such that

$$\mathcal{M}_i \in \mathcal{L}(V; V'), \quad |(\mathcal{M}_i \phi, \phi)| \leqq c|\phi| \, \|\phi\|.^2 \tag{2.10}$$

We consider the equation

$$\left(\frac{\partial y}{\partial t}, \phi\right) + a(y, \phi) - \sum_{i=1}^{m} v_i(t)(\mathcal{M}_i y, \phi) = (f, \phi), \tag{2.11}$$

$$y \in L^2(0, T; V), \quad \frac{\partial y}{\partial t} \in L^2(0, T; V'), \tag{2.12}$$

$$y(0) = y_0 \in H, \tag{2.13}$$

where $f \in L^2(0, T; V')$, $v_i \in L^2(0, T)$.

We prove first that under the hypothesis (4.12), Chap. 2, and (2.10), there exists a unique solution y of (2.11)–(2.13).

Everything relies on a priori estimates (for the way we can use these estimates, cf. Lions [7], Lions and Magenes [1], vol. 1, Chap. 3). If we take $\phi = y$ in (2.11),

---

[1] One has to prove the existence of such a y, under suitable hypotheses on the M_j's (cf. § 2.2 below).

[2] We recall that $|\cdot|$ (resp. $\|\cdot\|$) denotes the norm in H (resp. V).

we obtain, using (2.10),

$$\frac{1}{2}\frac{d}{dt}|y(t)|^2 + \alpha\|y(t)\|^2 - \lambda|y(t)|^2 - c\sum_{i=1}^{m}|v_i(t)|\,|y(t)|\,\|y(t)\|$$

$$\leqq \|f(t)\|_*\|y(t)\|\,;^3$$

hence,

$$\frac{1}{2}\frac{d}{dt}|y(t)|^2 + \alpha\|y(t)\|^2 \leqq \frac{\alpha}{2}\|y(t)\|^2 + c(1 + |v(t)|^2)|y(t)|^2 + c\|f(t)\|_*^2. \tag{2.14}$$

It follows that

$$|y(t)|^2 + \alpha\int_0^t \|y(\sigma)\|^2\,d\sigma \leqq c\int_0^t (1 + |v(\sigma)|^2)|y(\sigma)|^2\,d\sigma + c\int_0^t \|f(\sigma)\|_*^2 + |y_0|^2, \tag{2.15}$$

which, using Gronwall's inequality, implies that

$$|y(t)|^2 \leqq (|y_0|^2 + c\int_0^T \|f(\sigma)\|_*^2\,d\sigma)\exp c\int_0^T (1 + |v(\sigma)|^2)\,d\sigma = X \tag{2.16}$$

$$\int_0^T \|y(\sigma)\|^2\,d\sigma \leqq cX. \tag{2.17}$$

These estimates are sufficient for proving the existence as desired. The uniqueness is proved along the same lines.

Let us denote by $y(v)$ the solution of (2.11)–(2.13) and let us consider the cost function

$$J(v) = \int_0^T |y(t,v) - z_d(t)|^2\,dt + \sum_{j=1}^{m} N_j\int_0^T |v_j(t)|^2\,dt, \quad N_j > 0, \tag{2.18}$$

where z_d is given in $L^2(0, T; H)$. If U_{ad} denotes a closed convex subset of $(L^2(0, T))^m$, we consider the problem

$$\inf J(v), \quad v \in U_{\mathrm{ad}}. \tag{2.19}$$

We are going to prove an existence result, with a further hypothesis on $\mathcal{M}_i$. Let us introduce

$$W(0, T) = \{\phi|\phi \in L^2(0, T; V), \frac{\partial\phi}{\partial t} \in L^2(0, T; V')\}, \tag{2.20}$$

[3] $\|\cdot\|_*$ = norm in V'.

which is a Hilbert space for the norm

$$\left[\int_0^T \left(\|\phi(t)\|^2 + \left\|\frac{\partial\phi}{\partial t}(t)\right\|_*^2\right) dt\right]^{1/2}.$$

If $\phi \in W(0, T)$ we define $\mathscr{M}_i\phi$ by

$$(\mathscr{M}_i\phi)(t) = \mathscr{M}_i(\phi(t)). \tag{2.21}$$

We suppose that

$$\mathscr{M}_i \text{ is compact from } W(0, T) \to L^2(0, T; V'). \tag{2.22}$$

Then, *under the hypotheses* (2.10) *and* (2.22), *there exists* $u \in U_{\text{ad}}$ *such that*

$$J(u) \leqq J(v) \quad \textit{for all} \quad v \in U_{\text{ad}}. \tag{2.23}$$

Let indeed v^α be a minimizing sequence; v^α is then bounded in $(L^2(0, T))^m$; if we set $y(v^\alpha) = y^\alpha$, it follows from (2.17) that

$$y^\alpha \quad \text{remains in a bounded set of} \quad L^2(0, T; V). \tag{2.24}$$

Using (2.11) (with $v = v^\alpha$), we then obtain that

$$\frac{\partial y^\alpha}{\partial t} \quad \text{remains in a bounded set of} \quad L^2(0, T; V'); \tag{2.25}$$

but (2.24), (2.25) mean that y^α remains in a bounded set of $W(0, T)$. We can then extract a subsequence, still denoted by y^α, such that

$$y^\alpha \to y \quad \text{in } W(0, T) \text{ weakly}, \tag{2.26}$$

$$v^\alpha \to u \quad \text{in } (L^2(0, T))^m \text{ weakly}, \quad u \in U_{\text{ad}}. \tag{2.27}$$

But using the hypothesis (2.22), it follows that

$$\mathscr{M}_i y^\alpha \to \mathscr{M}_i y \quad \text{in } L^2(0, T; V') \text{ strongly} \tag{2.28}$$

and therefore,

$$v_i^\alpha \mathscr{M}_i y^\alpha \to u_i \mathscr{M}_i y \quad \text{in } L^1(0, T; V') \text{ weakly}. \tag{2.29}$$

Then we can pass to the limit in (2.10) (for $v = v^\alpha$) and we obtain $y = y(u)$. Then $\liminf J(v) \geqq J(u)$ and the result follows.

2.3. An example. We apply the result of § 2.2 to the situation of § 2.1, with $\mathscr{M}_j$ given by

$$\begin{aligned} &\mathscr{M}_j\phi = \frac{1}{|\sigma_j|}\int_{\sigma_j} \phi(x)\,dx, \quad \sigma_j \text{ bounded subset of } \Omega, \\ &|\sigma_j| = \text{measure of } \sigma_j. \end{aligned} \tag{2.30}$$

We take

$$V = H_0^1(\Omega), \quad H = L^2(\Omega), \quad a(\phi, \psi) = \int_\Omega \sum_{i=1}^n \frac{\partial \phi}{\partial x_i} \frac{\partial \psi}{\partial x_i} dx.$$

We can apply the results of § 2.2.

Since (2.10) follows directly, we have only to verify (2.22). Indeed if $\phi^\alpha \to \phi$ in $W(0, T)$ weakly, then, using a compactness result of Lions [2] Chap. 1, § 5.2, we have

$$\phi^\alpha \to \phi \quad \text{in } L^2(Q) \text{ strongly},$$

and then

$$\mathscr{M}_j\phi^\alpha(x, t) = \mathscr{M}_j\phi^\alpha(t) = \frac{1}{|\sigma_j|} \int_{\sigma_j} \phi(x, t)\, dx$$

converges in $L^2(Q)$ strongly (we assume that Ω is bounded).

3. Optimization with respect to the operators $\mathscr{M}_j$ of § 2.

3.1. The operators $\mathscr{M}_j$. Statement of the problem.

Notation. Let E be a closed subset of Ω, such that

(3.1) $$\text{distance } (E, \Gamma) \geqq r.$$

If $b \in E$, we set

(3.2) $$\sigma(b) = \text{ball with center } b \text{ and radius } r; \text{ volume of } \sigma(b) = \frac{1}{c}.$$

We then define $\mathscr{M}_j$ by:

(3.3) $$\mathscr{M}_j\phi = c \int_{\sigma(b_j)} \phi(x)\, dx,$$

which defines a continuous linear mapping from $L^2(\Omega) \to$ the constant functions ($cL^2(\Omega)$, Ω being bounded).

The state y of the system is given by the solution of (2.7), (2.2), (2.3), the $\mathscr{M}_j$'s being given by (3.3), with $b_j \in E$, $i \leqq j \leqq m$.

We allow now v *and* $b = \{b_1, \cdots, b_m\}$ to vary, subject to

(3.4) $$v \in U_{\text{ad}} = \text{closed convex subset of } (L^2(0, T))^m \ (= U),$$

(3.5) $$b \in E^m.$$

We denote the state by

$$y = y(x, t; v, b) = y(v, b).$$

The cost function is

(3.6) $$J(v, b) = \int_Q (y(x, t; v, b) - z_d(x, t))^2\, dx\, dt + \sum_{j=1}^m N_j \int_0^T v_j^2(t)\, dt, \quad N_j > 0.$$

We want to solve the problem:

(3.7) $$\inf J(v, b), \quad v \in U_{ad}, \quad b \in E^m.$$

We are going to show that there exists $\{u, b_0\}$ such that

(3.8) $$\begin{gathered} J(u, b_0) \leqq J(v, b) \quad \text{for all} \quad v \in U_{ad}, \quad \text{for all} \quad b \in E^m, \\ u \in U_{ad}, \quad b_0 \in E^m. \end{gathered}$$

3.2. Proof of (3.8). Optimal positions of the b_j's, inside Ω and on the boundary of Ω. Let v^α, b^α be a minimizing sequence. It follows from (3.6) that

(3.9) $$v^\alpha \text{ remains in a bounded set of } U.$$

We can easily check that the operators $\mathcal{M}_j$ defined by (3.3) satisfy (2.10) with constants c which do not depend on $b_j \in E$. It follows that $y^\alpha = y(v^\alpha, b^\alpha)$ satisfies (2.24), (2.25).

Therefore we can extract a subsequence, still denoted by v^α, y^α, b^α such that one has (2.26), (2.27) and

(3.10) $$b^\alpha \to b_0 \text{ in } (\mathbb{R}^n)^m, \quad b \in E^m.$$

The proof is completed as at the end of § 2.2, provided we prove that

(3.11) $$v_j^\alpha(t) \int_{\sigma(b_j^\alpha)} y^\alpha(x, t)\, dx \to u_j(t) \int_{\sigma(b_{0j})} y(x, t)\, dx \quad \text{in } L^1(Q) \text{ weakly}.$$

Let us drop the index j in (3.11). Since $v^\alpha \to u$ in $L^2(0, T)$ weakly, we have to prove that

(3.12) $$\int_{\sigma(b^\alpha)} y^\alpha(x, t)\, dx \to \int_{\sigma(b_0)} y(x, t)\, dx \quad \text{in } L^2(Q) \quad \text{(strongly)}.$$

But

(3.13) $$\int_{\sigma(b^\alpha)} y^\alpha\, dx - \int_{\sigma(b_0)} y\, dx = \int_{\sigma(b^\alpha)} (y^\alpha - y)\, dx + \int_{\sigma(b^\alpha) - \sigma(b_0)} y\, dx = g^\alpha + h^\alpha;$$

we have

$$|g^\alpha| \leqq \int_\Omega |y^\alpha - y|\, dx,$$

so that

$$\int_Q |g^\alpha|^2\, dx\, dt \to 0$$

since $y^\alpha \to y$ in $L^2(Q)$ *strongly* (cf. § 2.3).

We have also:

$$|h^\alpha| \leqq c \left(\int_\Omega y^2\, dx \right)^{1/2} |b^\alpha - b_0|;$$

hence,

$$\int_{\theta} |h^{\alpha}|^2 \,dx\,dt \leqq c|b^{\alpha} - b_0|^2 \to 0.$$

Thus (3.12) follows.

Remark 3.1. The result we have just proved shows the existence of optimal positions for the b_j's. This result can be extended, by similar methods:

(i) to systems governed by hyperbolic operators or Petrowsky operators;

(ii) to systems governed by nonlinear P.D.E.'s.

Remark 3.2. We assumed the b_j's to be "inside" Ω.

One can also treat the case where the b_j's are on Γ (see Fig. 5). For a given $b \in \Gamma$, let us define

(3.14) $$S_b = \sigma(b) \cap \Gamma$$

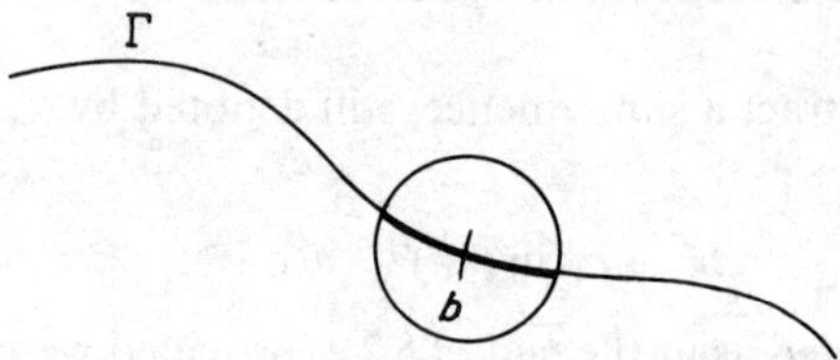

FIG. 5

(where $\sigma(b)$ is defined by (3.2)), and for $\phi \in H^1(\Omega)$ let us set

(3.15) $$\mathcal{N}_b(\phi) = \frac{1}{|S_b|}\int_{S_b} \phi \,d\Gamma,$$

where $|S_b|$ = area of S_b on Γ; (3.15) is well-defined since for all $\phi \in H^1(\Omega)$ we can define the trace of ϕ on Γ and $\phi \in H^{1/2}(\Gamma)$.[4]

Moreover,

(3.16) $$|\mathcal{N}_b(\phi)| \leqq c\left(\int_{\Gamma} |\phi|^2 \,dt\right)^{1/2} \leqq c|\phi|^{1/2}\|\phi\|^{1/2}$$ [5]

so that

(3.17) $$|(\mathcal{N}_b(\phi), \phi)| \leqq c|\phi|^{3/2}\|\phi\|^{1/2},$$

and the results of § 2.2 apply: given $b_1, \cdots, b_m \in \Gamma$, there exists a unique function y which satisfies

(3.18) $$\frac{\partial y}{\partial t} - \Delta y - \sum_{j=1}^{n} v_j(t)\mathcal{N}_{b_j}(y) = f \quad \text{in } Q,$$

[4] Cf. Lions and Magenes, [1], vol. 1, Chap. 1. We assume the boundary Γ to be smooth.

[5] Cf. Lions and Magenes, loc. cit.

with the conditions (2.2), (2.3) and

(3.19) $$y \in L^2(0, T; H_0^1(\Omega)), \quad \frac{\partial y}{\partial t} \in L^2(0, T; H^{-1}(\Omega)).$$

If we consider the cost functions given by (3.6), we have: there exists an optimal control u and an optimal set of positions $b_0 = \{b_{0j}\}$ on $\Gamma \times \cdots \times \Gamma$ such that

(3.20) $$J(u, b_0) \leqq J(v, b) \quad \text{for all} \quad v \in U_{\mathrm{ad}}, \quad \text{for all} \quad b \in \Gamma^m.$$

The proof follows the same lines as above, the crucial point being the following: *if* $y^\alpha \to y$ *in* $W(0, T)$ *weakly* ($W(0, T)$ is defined by (2.20)), *then* $y^\alpha \to y$ *in* $L^2(\Sigma)$ *strongly*.

Proof. By using the compactness result of Lions [2], Chap. 1, § 5.2 (already used in § 2.3) we see that

(3.21) $$y^\alpha \to y \quad \text{in } L^2(0, T; H^{1-\varepsilon}(\Omega)) \ \textit{strongly} \quad \text{for all} \quad \varepsilon > 0.$$

But let us choose in (3.21) $\varepsilon < \frac{1}{2}$; then the trace operator on Γ is continuous from $H^{1-\varepsilon}(\Omega) \to L^2(\Gamma)$ in particular, and (3.21) implies that $y^\alpha \to y$ in $L^2(\Sigma)$ strongly.

Remark 3.3. We can also introduce in $J(v, b)$ an extra term, say, $\Psi(b_1, \cdots, b_m)$, taking into account the implementation of the device at point b_j. This introduces no extra difficulty.

Remark 3.4. The optimal positions of the b_j's will in general depend on the z_d's. If we consider "desired" functions z_d of the form

(3.22) $$z_d(x, t) = \sum_{i=1}^{q} \lambda_i \theta_i(x, t), \quad \lambda_i \in \mathbb{R}, \quad \theta_i \text{ given in } L^2(Q),$$

then $J(v, b) = \mathscr{G}(v, b; \lambda_1, \cdots, \lambda_q)$ and we can introduce *as final cost function*:

(3.23) $$\hat{J}(v, b) = \int_{\mathbb{R}^q} \mathscr{G}(v, b; \lambda_1, \cdots, \lambda_q)\, d\mu(\lambda).$$

(More generally we could use cylindrical measures on the space spanned by z_d.)

4. Problems where one control variable is the geometrical domain.

4.1. Notation. State of the system. Let σ be a bounded open set in $\mathbb{R}^n$, Γ_0 a fixed part of its boundary (see Fig. 6) and let Ω_λ be a family of open sets satisfying (where $\lambda \in [0, 1]$)

(4.1) $$\Omega_\lambda \subset \sigma, \quad \partial\Omega_\lambda = \Gamma_0 \cup S_\lambda.$$

We suppose that the family Ω_λ "depends continuously on λ", in the following sense:

(4.2) $$\text{if} \quad \lambda \to \lambda_0, \quad \text{distance} \quad (S_\lambda, S_{\lambda_0}) \to 0.$$

Let us define

(4.3) $$Q = \sigma \times]0, T[, \quad Q_\lambda = \Omega_\lambda \times]0, T[.$$

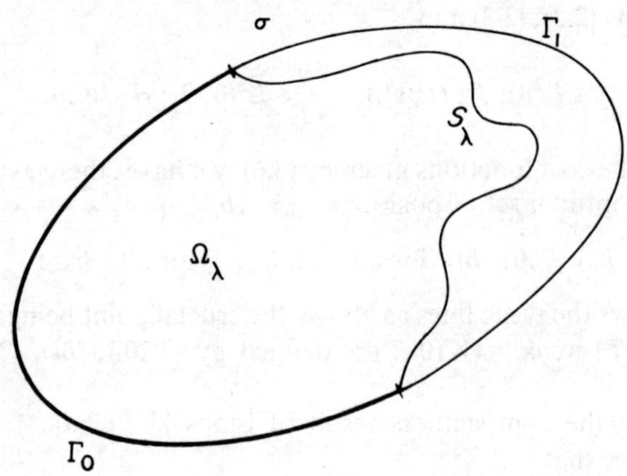

FIG. 6

The state of the system is given by the solution of

(4.4) $$\frac{\partial y}{\partial t} - \Delta y = f \quad \text{in } Q_\lambda,$$

where f is given in $L^2(Q)$,[6] with the boundary conditions

(4.5) $$\frac{\partial y}{\partial n} = v \quad \text{on} \quad \Sigma_0 = \Gamma_0 \times]0, T[,$$

where $v \in L^2(\Sigma_0)$, and

(4.6) $$y = 0 \quad \text{on} \quad \mathscr{S}_\lambda = S_\lambda \times]0, T[,$$

and the initial condition

(4.7) $$y(x, 0) = y_0(x) \quad \text{in } \Omega_\lambda,$$

where y_0 is given in $L^2(\sigma)$.[7]

This problem admits a unique solution that we shall denote by

(4.8) $$y = y(x, t; v, \lambda) = y(v, \lambda)$$

and which satisfies:

(4.9) $$y \in L^2(0, T; H^1(\Omega_\lambda)), \quad \frac{\partial y}{\partial t} \in L^2(0, T; H^{-1}(\Omega_\lambda)).$$

---

[6] So that the right-hand side of (4.4) actually denotes the *restriction* of f to Q_λ.

[7] So that the right-hand side of (4.7) actually denotes the *restriction* of y_0 to Ω_λ.

The variational formulation of the problem is as follows: for ϕ, ψ in $H^1(\Omega_\lambda)$ or in $H^1(\sigma)$, we shall set

$$a_\lambda(\phi, \psi) = \sum_{i=1}^{n} \int_{\Omega_\lambda} \frac{\partial \phi}{\partial x_i} \frac{\partial \psi}{\partial x_i} dx, \quad a(\phi, \psi) = \sum_{i=1}^{n} \int_{\sigma} \frac{\partial \phi}{\partial x_i} \frac{\partial \psi}{\partial x_i} dx, \tag{4.10}$$

$$(\phi, \psi)_\lambda = \int_{\Omega_\lambda} \phi\psi \, dx, \quad (\phi, \psi) = \int_{\sigma} \phi\psi \, dx; \tag{4.11}$$

let us define

$$V_\lambda = \{\phi | \phi \in H^1(\Omega_\lambda), \quad \phi = 0 \quad \text{on } S_\lambda\}; \tag{4.12}$$

then y is defined by

$$\left(\frac{\partial y}{\partial t}, \phi\right) + a_\lambda(y, \phi) = (f, \phi)_\lambda + \int_{\Gamma_0} v\phi \, d\Gamma_0 \quad \text{for all} \quad \phi \in V_\lambda, \tag{4.13}$$

$$\begin{gathered} y \in L^2(0, T; V_\lambda), \quad \frac{\partial y}{\partial t} \in L^2(0, T; V'_\lambda), \\ y(0) = y_0. \end{gathered} \tag{4.14}$$

4.2. Cost function. Existence of an optimal domain. The cost function is defined by

$$J(v, \lambda) = \int_{\Sigma_0} (y(v, \lambda) - z_d)^2 \, d\Sigma_0 + N \int_{\Sigma_0} v^2 \, d\Sigma_0, \quad N > 0. \tag{4.15}$$

Let U be $L^2(\Sigma_0)$ and let U_{ad} be a closed convex subset of U.

The problem is to find

$$\inf J(v, \lambda), \quad v \in U_{\text{ad}}, \quad \lambda \in [0, 1]. \tag{4.16}$$

We are going to prove the existence of $u \in U_{\text{ad}}, \lambda_0 \in [0, 1]$ such that

$$J(u, \lambda_0) \leqq J(v, \lambda) \quad \text{for all} \quad v \in U_{\text{ad}}, \quad \text{for all} \quad \lambda \in [0, 1]. \tag{4.17}$$

Let v_n, λ_n be a minimizing sequence. Due to the form of (4.15),

$$v_n \text{ remains in a bounded set of } U. \tag{4.18}$$

Let us set

$$y(v_n, \lambda_n) = y_n. \tag{4.19}$$

For $\phi \in V_\lambda$ let us denote by $\tilde{\phi}$ the extension of ϕ to σ by 0 outside Ω_λ; the mapping

$$\phi \to \tilde{\phi}$$

is continuous from $V_\lambda \to H^1(\sigma)$ and actually

$$\|\tilde{\phi}\|_{H^1(\sigma)} = \|\tilde{\phi}\| = \|\phi\|_{V_\lambda}.$$

Since $y(v,\lambda) = 0$ on S_λ, it follows from (4.13) that

$$(4.20)\qquad \left(\frac{\partial \tilde{y}_n}{\partial t}, \tilde{\phi}\right) + a(\tilde{y}_n, \tilde{\phi}) = (f, \tilde{\phi}) + \int_{\Gamma_0} v_n \tilde{\phi}\, d\Gamma_0 \quad \text{for all} \quad \tilde{\phi} \in \tilde{V}_{\lambda_n},$$

where $\tilde{V}_\lambda$ denotes the space spanned by $\tilde{\phi}$ when ϕ spans V_λ. But if we take $\tilde{\phi} = \tilde{y}_n$ in (4.20) we obtain

$$\frac{1}{2}\frac{d}{dt}|\tilde{y}_n|^2 + a(\tilde{y}_n, \tilde{y}_n) = (f, \tilde{y}_n) + \int_{\Gamma_0} v_n \tilde{y}_n\, d\Gamma_0$$

and using (4.18) it follows that

(4.21) $\tilde{y}_n$ remains in a bounded set of $L^2(0, T; H^1(\sigma)) \cap L^\infty(0, T; L^2(\sigma))$.

In order to be able to pass to the limit we need another estimate which uses *fractional t-derivatives.*[8]

Let us set

(4.22) $\tilde{y}_n(T) = \xi_n$ (which belongs to a bounded set of $L^2(\sigma)$)

and let us denote by $\hat{y}_n, \hat{f}, \hat{v}_n$ the Fourier transform in t of $\tilde{y}_n, f, v_n$ extended by 0 outside $(0, T)$; for instance,

$$(4.23)\qquad \hat{y}_n(\tau) = \int_0^T \tilde{y}_n(t)\exp(-it\tau)\, dt.$$

If we write (4.20) *on* $\mathbb{R}_t$ we obtain

$$(4.24)\qquad \left(\frac{\partial \tilde{y}_n}{\partial t}, \tilde{\phi}\right) + a(\tilde{y}_n, \tilde{\phi}) = (f, \tilde{\phi}) + \int_{\Gamma_0} v_n \tilde{\phi}\, d\Gamma_0 + (y_0, \tilde{\phi})\delta(t) - (\xi_n, \tilde{\phi})\delta(t - T).$$

Taking the Fourier transform in t of both sides of (4.24) we obtain

$$(4.25)\qquad i\tau(\hat{y}_n, \tilde{\phi}) + a(\hat{y}_n, \tilde{\phi}) = (\hat{f}, \tilde{\phi}) + \int_{\Gamma_0} \hat{v}_n \tilde{\phi}\, d\Gamma_0 + (y_0, \tilde{\phi}) - (\xi_n, \tilde{\phi})\exp(-i\tau T).$$

In (4.25) we choose $\tilde{\phi} = \hat{y}_n(\tau)$ and we take the absolute value of the imaginary parts of the corresponding equality. We obtain

$$(4.26)\qquad \begin{aligned} |\tau|\,|\hat{y}_n(\tau)|^2 &= \left|\operatorname{Im}\left\{(\hat{f}, \hat{y}_n) + \int_{\Gamma_0} \hat{v}_n \hat{y}_n\, d\Gamma_0 + (y_0, \hat{y}_n) \right.\right. \\ &\qquad \left.\left. - (\xi_n, \hat{y}_n)\exp(-i\tau T)\right\}\right| \\ &\leqq c[|\hat{y}_n(\tau)|\,|\hat{f}(\tau)| + |\hat{y}_n(\tau)| + \|\hat{v}_n(\tau)\|_{L^2(\Gamma_0)}|\hat{y}_n|^{1/2}\|\hat{y}_n\|^{1/2}]. \end{aligned}$$

Using (4.21) and (4.22) it follows from (4.26) that

$$(4.27)\qquad \int_{-\infty}^{+\infty} \frac{|\tau|}{1 + |\tau|^\beta}|\hat{y}_n(\tau)|^2\, d\tau \leqq c \quad \text{if} \quad \beta > \tfrac{1}{2},$$

[8] Cf. Lions [9], [2] for other applications of this method.

which expresses that

(4.28) $D^\gamma \tilde{y}_n$ remains in a bounded set of $L^2(0, T; L^2(\sigma))$ for $0 < \gamma < \frac{1}{4}$.

We can now prove (4.17). We can extract a subsequence, still denoted by v_n, λ_n, such that

$$v_n \to u \quad \text{in } L^2(\Sigma_0) \text{ weakly}, \quad u \in U_{ad}, \tag{4.29}$$

$$\lambda_n \to \lambda_0 \quad \text{in } [0, 1], \tag{4.30}$$

$$\tilde{y}_n \to Y \quad \text{in } L^2(0, T; H^1(\sigma)) \text{ weakly}, \tag{4.31}$$

$$D^\gamma \tilde{y}_n \to D^\gamma Y \quad \text{in } L^2(0, T; L^2(\sigma)) \text{ weakly}. \tag{4.32}$$

According to a compactness result given in Lions [9], [2], (4.31) and (4.32) imply that

$$\tilde{y}_n \to Y \quad \text{in } L^2(Q) \textit{ strongly}. \tag{4.33}$$

It remains to show that

$$Y = \tilde{y}(u, \lambda_0) \tag{4.34}$$

(it is then obvious that $\liminf J(v_n, \lambda_n) \geqq J(u, \lambda_0)$, hence the result follows).

We prove (4.34) in two steps:

$$Y = 0 \quad \text{outside } Q_{\lambda_0}; \tag{4.35}$$

the restriction y of Y to Q_{λ_0} satisfies

$$\left(\frac{\partial y}{\partial t}, \phi\right)_{\lambda_0} + a_{\lambda_0}(y, \psi) = (f, \phi)_{\lambda_0} + \int_{\Gamma_0} v\phi \, d\Gamma_0 \quad \text{for all} \quad \phi \in V_{\lambda_0}, \tag{4.36}$$

$$Y(0) = y_0.$$

Proof of (4.35). Let χ_λ = charăcteristic function of Ω_λ on σ.

It follows from (4.2) that $\chi_{\lambda_n} \to \chi_{\lambda_0}$ in $L^\infty(\sigma)$ weak star and, by virtue of (4.33), this implies that $\chi_{\lambda_n}\tilde{y}_n \to \chi_{\lambda_0} Y$ in $L^2(Q)$ weakly. But $\chi_{\lambda_n}\tilde{y}_n = \tilde{y}_n$ so that we obtain at the limit $\chi_{\lambda_0} Y = Y$; hence (4.35) follows.

This implies that $Y = 0$ on $S_{\lambda_0} \times]0, T[$.

Remark 4.1. Actually we could avoid using (4.33) here (hence (4.28)) but this proof shows that we can relax the hypothesis (4.2.).

Proof of (5.36). Let us consider a function $\psi(x, t)$ which satisfies

$$\psi \equiv 0 \text{ in a neighborhood of } S_{\lambda_0} \tag{4.37}$$

(the support of ψ for fixed t is represented by the shaded area on Fig. 7), and

$$\psi \in L^2(0, T; H^1(\Omega_{\lambda_0})), \quad \frac{\partial \phi}{\partial t} \in L^2(Q_{\lambda_0}), \quad \phi(x, T) = 0. \tag{4.38}$$

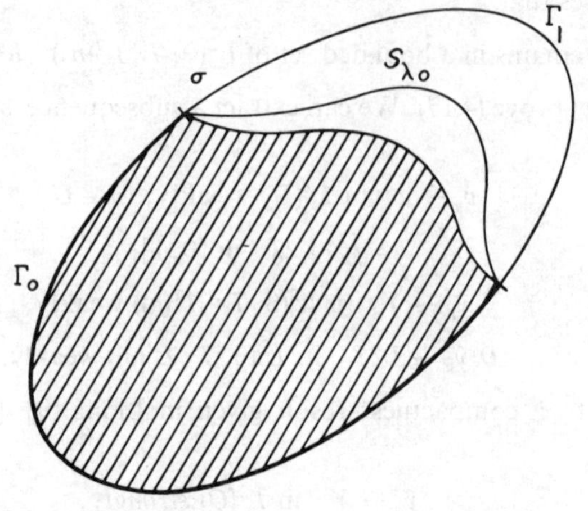

FIG. 7

It follows from (4.2) that in (4.20) we can choose, $\tilde{\psi}(x) = \phi(x, t)$ (since "$S_{\lambda_n} \to S_{\lambda_0}$", ψ is zero in a neighborhood of $S_\lambda \times]0, T[$). Then, integrating in t, we obtain

$$\int_{Q_{\lambda_0}} \left[-\tilde{y}_n \frac{\partial \phi}{\partial t} + \sum_{i=1}^{n} \frac{\partial \tilde{y}_n}{\partial x_i} \frac{\partial \phi}{\partial x_i} - f\phi \right] dx\, dt = \int_{\Sigma_0} v_n \phi \, d\Gamma_0 \, dt + \int_{\Omega_{\lambda_0}} y_0(x)\phi(x, 0)\, dx$$

and we can pass to the limit in this identity, to obtain

$$\int_{Q_{\lambda_0}} \left[-y \frac{\partial \phi}{\partial t} + \sum_{i=1}^{n} \frac{\partial y}{\partial x_i} \frac{\partial \phi}{\partial x_i} - f\phi \right] dx\, dt = \int_{\Sigma_0} u \phi \, d\Gamma_0 \, dt + \int_{\Omega_{\lambda_0}} y_0(x)\phi(x, 0)\, dx. \tag{4.39}$$

But the set of functions satisfying (4.37) is *dense* in the space defined by (4.39),[9] so that (4.39) is true for every ψ satisfying (4.38); hence (4.36) follows.

Remark 4.2. A physical problem leading to a situation of the preceding kind was indicated to us by T. L. Johnson (personal communication).

Another problem of a somewhat similar nature was indicated to us by I. Babuska (personal communication) (cf. J. L. Lions, Proceedings of the Fourth I.F.I.P. Colloquium of Optimization Techniques, Los Angeles, October, 1971).

---

[9] Provided with the norm $\left(\|\phi\|^2_{L^2(0,T;H^1(\Omega_{\lambda_0}))} + \left\| \frac{\partial \phi}{\partial t} \right\|_{L^2(Q_{\lambda_0})} \right)^{1/2}$.

CHAPTER 5

Singular Perturbations and Optimal Control

1. General remarks. Let us consider a system whose state $y = y_\varepsilon$ is given by the solution of

$$(\varepsilon\Lambda_1 + \Lambda_0)y_\varepsilon = f + Bv, \tag{1.1}$$

where, roughly speaking, Λ_1 is a P.D. operator of order greater than Λ_0, and where ε is a "small" positive parameter. Of course in (1.1) y_ε is subject to appropriate boundary conditions. Let the cost function be given by

$$J_\varepsilon(v) = \phi(y_\varepsilon(v)) + \psi(v). \tag{1.2}$$

If in (1.1) we take $\varepsilon = 0$, we obtain the *reduced system*

$$\Lambda_0 y = f + Bv, \tag{1.3}$$

where y is subject to corresponding *reduced* boundary conditions.

The cost function becomes

$$J(v) = \phi(y(v)) + \psi(v), \tag{1.4}$$

this at least if ϕ is defined on the (bigger) space described by y when v spans the set U_{ad} of admissible controls.

The problems which arise are the following:

(i) to justify the preceding procedure, in specific situations;

(ii) to ascertain whether it is possible to obtain better approximations of (1.1)–(1.2) than of problem (1.3)–(1.4).

Without trying here to cover completely these problems (cf. other results and problems in Lions [12]) we are going to give some examples.

2. An elliptic system.

2.1. The state of the system. We suppose that the state $\{y_{1\varepsilon}, y_{2\varepsilon}\}$ of the system is given by

$$\begin{aligned} -\Delta y_{1\varepsilon} + y_{1\varepsilon} + y_{2\varepsilon} &= f_1 + v_1, \\ -\varepsilon\Delta y_{2\varepsilon} + y_{2\varepsilon} - y_{1\varepsilon} &= f_2 + v_2 \quad \text{in } \Omega, \qquad \varepsilon > 0, \end{aligned} \tag{2.1}$$

subject to the boundary conditions:

$$y_{1\varepsilon} = 0, \quad y_{2\varepsilon} = 0 \quad \text{on } \Gamma. \tag{2.2}$$

In (2.1) we suppose that $f_1, f_2 \in L^2(\Omega)$ and that

$$v = \{v_1, v_2\} \in U_{\mathrm{ad}} = \text{closed convex subset of } U,\ U = (L^2(\Omega))^2. \tag{2.3}$$

The problem (2.1)–(2.2) uniquely defines $\{y_{1\varepsilon}, y_{2\varepsilon}\}$. If we set

$$a(\phi, \psi) = \int_\Omega \sum_{i=1}^n \frac{\partial \phi}{\partial x_i} \frac{\partial \psi}{\partial x_i} dx, \tag{2.4}$$

the variational formulation of the problem is:

$$\begin{aligned} &a(y_{1\varepsilon}, \phi_1) + (y_{1\varepsilon} + y_{2\varepsilon}, \phi_1) + \varepsilon a(y_{2\varepsilon}, \phi_2) + (y_{2\varepsilon} - y_{1\varepsilon}, \phi_2) \\ &\qquad = (f_1 + v_1, \phi_1) + (f_2 + v_2, \phi_2) \quad \text{for all} \quad \phi_i \in H_0^1(\Omega), \end{aligned} \tag{2.5}$$

where $\{y_{1\varepsilon}, y_{2\varepsilon}\} \in H_0^1(\Omega) \times H_0^1(\Omega)$.

If we take $\phi_i = y_{i\varepsilon}$, $i = 1, 2$, we obtain

$$a(y_{1\varepsilon}, y_{1\varepsilon}) + \varepsilon a(y_{2\varepsilon}, y_{2\varepsilon}) + |y_{1\varepsilon}|^2 + |y_{2\varepsilon}|^2 = (f_1 + v_1, y_{1\varepsilon}) + (f_2 + v_2, y_{2\varepsilon}); \tag{2.6}$$

hence it follows that, when $\varepsilon \to 0$:

$$\|y_{1\varepsilon}\|^2_{H_0^1(\Omega)} + \|y_{2\varepsilon}\|^2_{L^2(\Omega)} \leq C, \tag{2.7}$$

$$\sqrt{\varepsilon} y_{2\varepsilon} \quad \text{remains in a bounded set of} \quad H_0^1(\Omega). \tag{2.8}$$

We can easily obtain from these estimates that, when $\varepsilon \to 0$,

$$\begin{aligned} y_{1\varepsilon} &\to y_1 \quad \text{in } H_0^1(\Omega), \\ y_{2\varepsilon} &\to y_2 \quad \text{in } L^2(\Omega), \end{aligned} \tag{2.9}$$

where $\{y_1, y_2\}$ is the solution of the reduced system

$$\begin{aligned} -\Delta y_1 + y_1 + y_2 &= f_1 + v_1, \\ y_2 - y_1 &= f_2 + v_2, \end{aligned} \tag{2.10}$$

with the only boundary condition

$$y_1 = 0 \quad \text{on } \Gamma. \tag{2.11}$$

Of course we can eliminate y_2 in (2.10); hence

$$-\Delta y_1 + 2y_1 = f_1 - f_2 + v_1 - v_2. \tag{2.12}$$

2.2. Cost function for the system and for the reduced system. We consider the cost function given by

$$\begin{aligned} J_\varepsilon(v) = \int_\Omega [|y_{1\varepsilon} - z_{1d}|^2 + |y_{2\varepsilon} - z_{2d}|^2]\, dx + N_1 \int_\Omega v_1^2\, dx + N_2 \int_\Omega v_1^2\, dx, \\ N_1, N_2 > 0. \end{aligned} \tag{2.13}$$

The reduced cost function is given by

$$\begin{aligned} &J(v) = \int_\Omega [|y_1 - z_{1d}|^2 + |y_2 - z_{2d}|^2]\, dx + N_1 \int_\Omega v_1^2\, dx + N_2 \int_\Omega v_2^2\, dx, \\ &y_2 = y_1 + f_2 + v_2. \end{aligned} \tag{2.14}$$

Let us denote by u_ε (resp. u) the optimal control for the problem (resp. for the reduced problem). Then we have:

(2.15) when $\varepsilon \to 0$, $u_\varepsilon \to u$ in $L^2(\Omega) \times L^2(\Omega)$ *weakly*, and $J_\varepsilon(u_\varepsilon) \to J(u)$.

Moreover, if we set

(2.16) $$y_\varepsilon(u_\varepsilon) = \{y_{\varepsilon_1}, y_{\varepsilon_2}\}, \quad y(u) = y = \{y_1, y_2\},$$

then

(2.17) $$y_\varepsilon \to y \quad \text{in } H_0^1(\Omega) \times L^2(\Omega) \text{ weakly}.$$

Indeed, it follows from (2.9) that, if $\varepsilon \to 0$, then

(2.18) $$J_\varepsilon(v) \to J(v) \quad \text{for all} \quad v \in U.$$

Therefore,

$$J_\varepsilon(u_\varepsilon) \leqq J_\varepsilon(v_0) \leqq \text{const.}$$

and since

$$J_\varepsilon(v) \geqq \inf(N_1, N_2) \int_\Omega (v_1^2 + v_2^2)\, dx,$$

it follows that

(2.19) u_ε remains in a bounded subset of U_{ad}.

But this implies that

(2.20) $y_\varepsilon(u_\varepsilon) = y_\varepsilon$ remains in a bounded set of $H_0^1(\Omega) \times L^2(\Omega)$.

Therefore we can extract a subsequence, still denoted by u_ε, y_ε, such that

(2.21) $$\begin{aligned} &u_\varepsilon \to u \quad \text{in } U \text{ weakly}, \quad u \in U_{\text{ad}}, \\ &y_\varepsilon \to y \quad \text{in } H_0^1(\Omega) \times L^2(\Omega) \text{ weakly}. \end{aligned}$$

It follows from (2.21) and (2.1) (when $v_i = u_{i\varepsilon}$) that

(2.22) $$y = y(u).$$

Therefore, $\liminf J_\varepsilon(u_\varepsilon) \geqq J(u)$. But using (2.18) we obtain

$$\limsup J_\varepsilon(u_\varepsilon) \leqq J(v) \quad \text{for all} \quad v \in U_{\text{ad}};$$

hence $J(u) \leqq J(v)$ for all $v \in U_{\text{ad}}$. Thus the result follows.

Remark 2.1. The interest of the preceding result is of course to replace, for, say, computational purposes, the complete system by the reduced one.

The next question is now (ii), § 1. We give an example of such a question in the following section.

3. Use of a boundary layer.

3.1. Setting of the problem. We consider the domain $\Omega = \{x | x_n > 0\}$, we set $x' = \{x_1, \cdots, x_{n-1}\}$. We suppose that the state of the system is given by $y_\varepsilon(v)$, the solution of

$$-\varepsilon \Delta y_\varepsilon(v) + y_\varepsilon(v) = v(x')g(x_n) \quad \text{in } \Omega, \tag{3.1}$$

$$y_\varepsilon(x', 0; v) = 0, \tag{3.2}$$

where g is given satisfying

$$g \in H^1(0, \infty), \quad g(0) = 1. \tag{3.3}$$

We take

$$U = H^1(\mathbb{R}^{n-1}_{x'}) \tag{3.4}$$

and in (3.1) we suppose that

$$v \in U_{\text{ad}} = \text{closed convex subset of } U. \tag{3.5}$$

The cost function we consider is given by

$$J_\varepsilon(v) = |y_\varepsilon(v) - z_d|^2 + N\|v\|_U^2, \quad N > 0, \tag{3.6}$$

where z_d is given in $L^2(\Omega)$.

Let u_ε be the optimal control: $J_\varepsilon(u_\varepsilon) = \inf J_\varepsilon(v)$, $v \in U_{\text{ad}}$. When $\varepsilon \to 0$, $\|u_\varepsilon\|_U \leqq$ const. and

$$J_\varepsilon(u_\varepsilon) \to \inf J(v), \tag{3.7}$$

where

$$J(v) = |y(v) - z_d|^2 + N\|v\|_U^2 \tag{3.8}$$

and

$$y(v) = vg. \tag{3.9}$$

We want to obtain another simple functional giving a better approximation of J_ε. For this purpose, we use a part of boundary layer theory.

3.2. Correctors. The corrected functional. Since $v \in U$ and g is given satisfying (3.3), we have

$$y(v) \in H^1(\Omega), \tag{3.10}$$

but $y_\varepsilon(v) \nrightarrow y(v)$ in $H^1(\Omega)$ if $v \neq 0$ since otherwise $y(v)|_\Gamma = 0$, which is absurd $(y(v)|_\Gamma = v)$. But if we introduce the "zero order corrector" $\theta_\varepsilon(v)$ given by

$$\theta_\varepsilon(v) = -v(x') \exp\left(-\frac{x_n}{\sqrt{\varepsilon}}\right), \tag{3.11}$$

then we have (cf. Visik and Lyusternik [1], Lions [10], [11])

$$y_\varepsilon(v) - (y(v) + \theta_\varepsilon(v)) \to 0 \quad \text{in } H^1(\Omega) \tag{3.12}$$

and

$$|y_\varepsilon(v) - (y(v) + \theta_\varepsilon(v))| \leqq C\sqrt{\varepsilon} \quad \text{uniformly for } v \in \text{bounded set of } U. \tag{3.13}$$

We now introduce the *corrected functional*:

$$G_\varepsilon(v) = |y(v) + \theta_\varepsilon(v) - z_d|^2 + N\|v\|_U^2. \tag{3.14}$$

Remark 3.1. One can introduce correctors of any order, and, therefore, corrected functionals of any order.

Let $\bar{u}_\varepsilon$ be the optimal control for G_ε:

$$G_\varepsilon(\bar{u}_\varepsilon) = \inf G_\varepsilon(v), \quad v \in U_{\mathrm{ad}}. \tag{3.15}$$

We have

$$|J_\varepsilon(v_\varepsilon) - G_\varepsilon(\bar{v}_\varepsilon)| \leqq C\sqrt{\varepsilon}. \tag{3.16}$$

Indeed, we know that $\|u_\varepsilon\|_U \leqq C$ and, in the same way, $\|\bar{u}_\varepsilon\|_U \leqq C$. Then, if we denote by $O_1(\sqrt{\varepsilon})$ any element of $L^2(\Omega)$ of norm $\leqq C\sqrt{\varepsilon}$, we have:

$$\begin{aligned} G_\varepsilon(u_\varepsilon) &= |y(u_\varepsilon) + \theta_\varepsilon(u_\varepsilon) - z_d|^2 + N\|u_\varepsilon\|_U^2 \qquad \text{(using (3.13))} \\ &= |y_\varepsilon(u_\varepsilon) - z_d + O_1(\sqrt{\varepsilon})|^2 + N\|u_\varepsilon\|_U^2 \\ &= J_\varepsilon(u_\varepsilon) + O(\sqrt{\varepsilon}) \end{aligned}$$

and, in the same manner

$$J_\varepsilon(\bar{u}_\varepsilon) = G_\varepsilon(\bar{u}_\varepsilon) + O(\sqrt{\varepsilon});$$

hence (3.16) follows.

Remark 3.2. Using higher order correctors, and the corrected functional, we can obtain a result similar to (3.16) but with any power of ε in the right-hand side.

4. An evolution system.

4.1. Setting of the problem. We use the notation of Chap. 2, § 4. We consider

$$A \in \mathscr{L}(V \times V; V')\ ^1 \tag{4.1}$$

such that

$$(A\phi, \phi) \geqq \alpha\|\phi\|^2 - \lambda|\phi|^2, \quad \alpha > 0, \quad \lambda \in \mathbb{R},$$

where $\phi = \{\phi_1, \phi_2\} \in V \times V$,

$$\|\phi\|^2 = \|\phi_1\|^2 + \|\phi_2\|^2, \quad |\phi|^2 = |\phi_1|^2 + |\phi_2|^2.$$

---

[1] We could as well consider $A \in \mathscr{L}(V_1 \times V_2; V_1' \times V_2')$, where $V_i \subset H_i \subset V_i'$, $i = 1, 2$, V_1 and V_2 being *distinct* Hilbert spaces.

We introduce, with I = identity,

$$D_\varepsilon = \begin{pmatrix} I & 0 \\ 0 & \varepsilon I \end{pmatrix}, \quad \varepsilon > 0, \tag{4.2}$$

and

$$B \in \mathscr{L}(\hat{U}; V' \times V'). \tag{4.3}$$

The state of the system is given by $y_\varepsilon(v)$, the solution of

$$D_\varepsilon \frac{\partial y_\varepsilon}{\partial t} + Ay_\varepsilon = f + Bv, \tag{4.4}$$

where f is given in $L^2(0, T; V' \times V')$,

$$y_\varepsilon(0) = y_0, \quad y_0 \text{ given in } H \times H. \tag{4.5}$$

The problem (4.4)–(4.5) admits a unique solution (same result as in Chap. 2, § 4).

Remark 4.1. The condition (4.5) can be equivalently written

$$D_\varepsilon y_\varepsilon(0) = D_\varepsilon y_0. \tag{4.6}$$

The cost function is given by

$$J_\varepsilon(v) = \int_0^T \|Cy_\varepsilon(v) - z_d\|_H^2 \, dt + \int_0^T (Nv, v)_{\hat{U}} \, dt. \tag{4.7}$$

We denote by u_ε the optimal control:

$$J_\varepsilon(u_\varepsilon) = \inf J_\varepsilon(v), \quad v \in U_{\text{ad}}, \quad U_{\text{ad}} = \text{closed convex subset of } U. \tag{4.8}$$

Remark 4.2. If $U_{\text{ad}} = U$, the optimal control is given by

$$B^* p_\varepsilon + Nu_\varepsilon = 0, \tag{4.9}$$

where the $\{y_\varepsilon, p_\varepsilon\}$ are given by

$$\begin{gathered} D_\varepsilon \frac{\partial y_\varepsilon}{\partial t} + Ay_\varepsilon + BN^{-1}B^* p_\varepsilon = f, \\ -D_\varepsilon \frac{\partial p_\varepsilon}{\partial t} + A^* p_\varepsilon - C^* C y_\varepsilon = -C^* z_d, \\ y_\varepsilon(0) = y_0, \quad p_\varepsilon(T) = 0, \\ y_\varepsilon, p_\varepsilon \in L^2(0, T; V \times V). \end{gathered} \tag{4.10}$$

We now want to let $\varepsilon \to 0$.

4.2. The reduced problem. We notice that

$$D_0 = \begin{pmatrix} I & 0 \\ 0 & 0 \end{pmatrix}. \tag{4.11}$$

The reduced state equation is given by

$$(4.12)\qquad \begin{aligned} &D_0\frac{\partial y}{\partial t} + Ay = f + Bv,\\ &D_0 y(0) = D_0 y_0. \end{aligned}$$

If we write $A = \begin{pmatrix} A_{11} & A_{12} \\ A_{21} & A_{22} \end{pmatrix}$, then (4.12) is equivalent to

$$(4.13)\qquad \begin{aligned} &\frac{\partial y_1}{\partial t} + A_{11}y_1 + A_{12}y_2 = (f + Bv)_1,\\ &A_{21}y_1 + A_{22}y_2 = (f + Bv)_2, \end{aligned}$$

with

$$(4.14)\qquad y_1(0) = y_{01}.$$

Let us suppose that

$$(4.15)\qquad (A\phi, \phi) \geqq \alpha\|\phi\|^2.$$

Then A_{22} is invertible, so that we can express y_2 by

$$(4.16)\qquad y_2 = -A_{22}^{-1}A_{21}y_1 + A_{22}^{-1}(f + Bv)_2$$

and we obtain for y_1 the equation

$$(4.17)\qquad \frac{\partial y_1}{\partial t} + A_{11}y_1 - A_{12}A_{22}^{-1}A_{21}y_1 = (f + Bv)_1 - A_{12}A_{22}^{-1}(f + Bv)_2.$$

If we suppose that[2]

$$(4.18)\qquad A_{12}A_{22}^{-1}A_{21} \in \mathscr{L}(V; H),$$

then (4.17), (4.14) admits a unique solution. Summing up, if we assume that (4.15) and (4.18) take place, then the reduced state equation (4.12) admits a unique solution, which satisfies:

$$y \in L^2(0, T; V \times V), \quad D_0\frac{\partial y}{\partial t} \in L^2(0, T; V' \times V').$$

The reduced optimal control problem is now to find

$$(4.19)\qquad \inf J(v), \quad v \in U_{\mathrm{ad}},$$

where

$$(4.20)\qquad J(v) = \int_0^T \|Cy(v) - z_d\|_H^2\,dt + \int_0^T (Nv, v)_{\mathscr{U}}\,dt.$$

[2] We can relax this hypothesis.

We are now going to check that, when $\varepsilon \to 0$,

$$J_\varepsilon(u_\varepsilon) \to \inf J(v) = J(u), \tag{4.21}$$

$$y_\varepsilon(u_\varepsilon) \to y(u) \quad \text{in } L^2(0, T; V \times V) \text{ weakly}, \tag{4.22}$$

$$u_\varepsilon \to u \quad \text{in } U \text{ weakly}. \tag{4.23}$$

The proof is along the same lines as in § 2.2; everything rests on the property:

$$y_\varepsilon(v) \to y(v) \quad (\text{when } \varepsilon \to 0) \quad \text{in } L^2(0, T; V \times V). \tag{4.24}$$

For proving (4.24), we first observe that it follows from (4.4) and (4.15) that (writing y_ε for $y_\varepsilon(v)$):

$$\frac{1}{2}\frac{d}{dt}(D_\varepsilon y_\varepsilon, y_\varepsilon) + \alpha \|y_\varepsilon\|^2 \leqq \|f + Bv\|_* \|y_\varepsilon\|;$$

hence,

$$\begin{aligned} &(D_\varepsilon y_\varepsilon(t), y_\varepsilon(t)) + \alpha \int_0^t \|y_\varepsilon(\sigma)\|^2 \, d\sigma \\ &\qquad \leqq C \int_0^t \|f + Bv\|_*^2 \, d\sigma + (D_\varepsilon y_0, y_0). \end{aligned} \tag{4.25}$$

Therefore,

$$y_\varepsilon \in \text{bounded set of } L^2(0, T; V \times V), \tag{4.26}$$

$$y_{\varepsilon 1} (\text{resp. } \sqrt{\varepsilon} y_{\varepsilon 2}) \text{ remains in a bounded set of } L^\infty(0, T; H). \tag{4.27}$$

Consequently, we can extract a subsequence, still denoted by y_ε, such that $y_\varepsilon \to y$ in $L^2(0, T; V \times V)$ weakly.

We can easily check that y is the solution of (4.12), so that

$$y = y(v). \tag{4.28}$$

It remains to prove that we have *strong* convergence in $L^2(0, T; V \times V)$. We consider (writing ϕ' for $\partial\phi/\partial t$):

$$\begin{aligned} X_\varepsilon &= \int_0^T (D_0(y'_\varepsilon - y'), y_\varepsilon - y) \, dt + \int_0^T ((D_\varepsilon - D_0) y'_\varepsilon, y_\varepsilon) \, dt \\ &\quad + \int_0^T (A(y_\varepsilon - y), y_\varepsilon - y) \, dt \\ &= \int_0^T (f + Bv, y_\varepsilon) \, dt - \int_0^T (D_0 y', y_\varepsilon - y) \, dt - \int_0^T (D_0 y'_\varepsilon, y) \, dt \\ &\quad - \int_0^T [(Ay_\varepsilon, y) + (Ay, y_\varepsilon)] \, dt + \int_0^T (Ay, y) \, dt; \end{aligned} \tag{4.29}$$

but $D_0 y'_\varepsilon \to D_0 y'$ in $L^2(0, T; V' \times V')$, so that (4.29) gives:

$$X_\varepsilon \to \int_0^T (f + Bv, y)\,dt - \int_0^T (D_0 y', y)\,dt - \int_0^T (Ay, y)\,dt = 0.$$

Then

$$X_\varepsilon + \tfrac{1}{2}((D_\varepsilon - D_0)y_0, y_0) = Y_\varepsilon \to 0. \tag{4.30}$$

But

$$Y_\varepsilon = \tfrac{1}{2}(D_0(y_\varepsilon(T) - y(T)), y_\varepsilon(T) - y(T)) + \tfrac{1}{2}((D_\varepsilon - D_0)y_\varepsilon(T), y_\varepsilon(T))$$
$$+ \int_0^T (A(y_\varepsilon - y), y_\varepsilon - y)\,dt \geqq \alpha \int_0^T \|y_\varepsilon - y\|^2\,dt$$

so that (4.30) shows (4.24).

4.3. The case without constraints. The above considerations apply to the case without constraints.

The reduced optimality system is given by

$$\begin{aligned} &D_0 \frac{\partial y}{\partial t} + Ay + BN^{-1}B^*p = f, \\ &-D_0 \frac{\partial p}{\partial t} + A^*p - C^*Cy = -C^*z_d, \\ &D_0 y(0) = D_0 y_0, \quad D_0 p(T) = 0, \\ &y, p \in L^2(0, T; V \times V). \end{aligned} \tag{4.31}$$

If $\{y_\varepsilon, p_\varepsilon\}$ denotes the solution of (4.10), we have

$$\{y_\varepsilon, p_\varepsilon\} \to \{y, p\} \text{ in } L^2(0, T; V \times V) \times L^2(0, T; V \times V). \tag{4.32}$$

But this gives in turn a result on the nonlinear P.D.E. (of Riccati type—cf. Chap. 2) corresponding to the uncoupling of (4.10) and of (4.31). We have

$$\begin{aligned} p_\varepsilon &= P_\varepsilon y_\varepsilon + r_\varepsilon, \\ p &= Py + r, \end{aligned} \tag{4.33}$$

where P_ε (resp. P) is the solution of

$$-D_\varepsilon \frac{\partial P_\varepsilon}{\partial t} + P_\varepsilon A + A^*P_\varepsilon + P_\varepsilon BN^{-1}B^*P = C^*C, \tag{4.34}$$

$$P_\varepsilon(T) = 0, \quad (P_\varepsilon(t)\phi, \psi) = (\phi, P_\varepsilon(t)\psi), \quad (P_\varepsilon(t)\phi, \phi) \geqq 0, \tag{4.35}$$

(resp. of

$$-D_0\frac{\partial P}{\partial t} + PA + A^*P + PBN^{-1}B^*P = C^*C, \tag{4.36}$$

$$P(T) = 0, \quad (P(t)\phi, \psi) = (\phi, P(t)\psi), \quad (P(t)\phi, \phi) \geqq 0\,). \tag{4.37}$$

We obtain that $P_\varepsilon \to P$ in the sense (for instance) that

$$P_\varepsilon(t)h \to P(t)h \quad \text{for all} \quad h \in H \times H. \tag{4.38}$$

(We do not know if there is in general uniqueness of the solution of (4.36)–(4.37).)

Remark 4.3. We do not know a direct proof of (4.35).

Remark 4.4. Using correctors, we can obtain asymptotic expansions for P_ε; we shall return to this problem in Lions [12].

4.4. The case when v depends only on t. We now use considerations analogous to those of § 4.3 but for the system

$$D_\varepsilon\frac{\partial y_\varepsilon}{\partial t} + Ay_\varepsilon = f + v(t)w, \tag{4.39}$$

where

$$v \in L^2(0, T) \tag{4.40}$$

and where $w \in V \times V$ satisfies

$$Aw = \lambda w. \tag{4.41}$$

Then we see, as in Chap. 2, § 7, that the optimality system relative to the cost function

$$J_\varepsilon(v) = \int_0^T \|Cy_\varepsilon - z_d\|_H^2\, dt + N\int_0^T v(t)^2\, dt \tag{4.42}$$

without constraints is given by:

$$\begin{aligned} &D_\varepsilon\frac{\partial y_\varepsilon}{\partial t} + Ay_\varepsilon + \frac{1}{N}(p_\varepsilon, w)w = f,\\ &-D_\varepsilon\frac{\partial p}{\partial t} + A^*p_\varepsilon - C^*Cy_\varepsilon = -C^*z_d,\\ &y_\varepsilon(0) = y_0, \quad p_\varepsilon(T) = 0. \end{aligned} \tag{4.43}$$

The reduced optimality system is given by

$$\begin{aligned} &D_0\frac{\partial y}{\partial t} + Ay + \frac{1}{N}(p, w)w = f,\\ &-D_0\frac{\partial p}{\partial t} + A^*p - C^*Cy = -C^*z_d,\\ &D_0y(0) = D_0y_0, \quad D_0p(T) = 0. \end{aligned} \tag{4.44}$$

The result (4.38) applies. If we set

$$P_\varepsilon w = Q_\varepsilon, \quad Pw = Q, \tag{4.45}$$

we have:

$$\begin{aligned} &-D_\varepsilon \frac{\partial Q_\varepsilon}{\partial t} + (A^* + \lambda)Q_\varepsilon + \frac{1}{N}(Q_\varepsilon, w)Q_\varepsilon = C^*Cw, \\ &Q_\varepsilon(T) = 0, \quad (Q_\varepsilon, w) \geqq 0, \end{aligned} \tag{4.46}$$

and

$$\begin{aligned} &-D_0 \frac{\partial Q}{\partial t} + (A^* + \lambda)Q + \frac{1}{N}(Q, w)Q = C^*Cw, \\ &Q(T) = 0, \quad (Q, w) \geqq 0. \end{aligned} \tag{4.47}$$

We have obtained that

$$Q_\varepsilon \to Q \quad \text{in } L^2(0, T; V \times V) \text{ weakly}. \tag{4.48}$$

Example 4.1. Let us consider the system whose state is given by

$$\begin{aligned} &\frac{\partial y_{\varepsilon 1}}{\partial t} - \Delta y_{\varepsilon 1} + \alpha y_{\varepsilon 2} = vw_1, \\ &\varepsilon \frac{\partial y_{\varepsilon 2}}{\partial t} - \Delta y_{\varepsilon 2} + \alpha y_{\varepsilon 1} = vw_1 \quad \text{in } Q, \\ &y_{\varepsilon i} = 0 \quad \text{on } \Sigma, \\ &y_{\varepsilon 1}(0) = y_{01}, \quad y_{\varepsilon 2}(0) = y_{02}, \end{aligned} \tag{4.49}$$

$$-\Delta w_1 = \mu w_1, \quad w_1 \in H_0^1(\Omega). \tag{4.50}$$

We apply the preceding remarks with $V = H_0^1(\Omega)$,

$$A = \begin{pmatrix} -\Delta & \alpha I \\ \alpha I & -\Delta \end{pmatrix}, \lambda = \mu + \alpha, C = \text{operator of mean value}$$

(see (7.35), Chap. 2). The equations which correspond to (4.46) (resp. (4.47)) are:

$$\begin{aligned} &-\frac{\partial Q_{\varepsilon 1}}{\partial t} - \Delta Q_{\varepsilon 1} + \alpha Q_{\varepsilon 2} + \frac{1}{N}\left(\int_\Omega (Q_{\varepsilon 1} + Q_{\varepsilon 2}) w_1 \, dx\right) Q_{\varepsilon 1} = \bar{w}_1, \\ &-\varepsilon \frac{\partial Q_{\varepsilon 2}}{\partial t} - \Delta Q_{\varepsilon 2} + \alpha Q_{\varepsilon 1} + \frac{1}{N}\left(\int_\Omega (Q_{\varepsilon 1} + Q_{\varepsilon 2}) w_1 \, dx\right) Q_{\varepsilon 2} = \bar{w}_1, \\ &Q_{\varepsilon i} = 0 \quad \text{on } \Sigma, \quad i = 1, 2, \\ &Q_{\varepsilon i}(x, T) = 0 \text{ in } \Omega, \quad \int_\Omega (Q_{\varepsilon 1} + Q_{\varepsilon 2}) w_1 \, dx \geqq 0, \end{aligned} \tag{4.51}$$

(resp.

$$
(4.52)\quad
\begin{aligned}
&-\frac{\partial Q_1}{\partial t} - \Delta Q_1 + \alpha Q_2 + \frac{1}{N}\left(\int_\Omega (Q_1 + Q_2)w_1\,dx\right)Q_1 = \bar{w}_1,\\
&-\Delta Q_2 + \alpha Q_1 + \frac{1}{N}\left(\int_\Omega (Q_1 + Q_2)w_1\,dx\right)Q_2 = \bar{w}_1,\\
&Q_i = 0 \quad \text{on } \Sigma, \quad i = 1, 2,\\
&Q_i(x, T) = 0 \quad \text{in } \Omega, \quad \int_\Omega (Q_1 + Q_2)w_1\,dx \geqq 0\ ).
\end{aligned}
$$

We have thus shown that *the* solution $Q_\varepsilon = \{Q_{\varepsilon 1}, Q_{\varepsilon 2}\}$ of (4.51) converges as $\varepsilon \to 0$, toward *a* solution Q of (4.52) (we do not know if the solution $Q \in L^2(0, T; H_0^1(\Omega) \times H_0^1(\Omega))$ of (4.52) is unique).

4.5. Various remarks. 1. In the case of nonlinear lumped parameter systems, consideration of the type of those of §§ 4.1, 4.2 have been introduced in Kokotovic and Sannuti [1], Kokotovic and Yackel [1], Sannuti and Kokotovic [1], [2], R. E. O'Malley [1], [2].

2. Similar considerations apply to hyperbolic systems (or Petrowski systems):

$$
D_\varepsilon \frac{\partial^2 y_\varepsilon}{\partial t^2} + A y_\varepsilon = f + Bv.
$$

CHAPTER 6

Remarks on Numerical Methods

1. Discretization.

1.1. Approximation of the state equation. Since the state equations we consider here are P.D.E. (linear or nonlinear), the approximation theory of the state equation is already a non-negligible part of *all* numerical analysis! We can only here confine ourselves to simple indications.

To start with, we consider, in the notation of Chap. 2, § 4.1, the state y given by

$$\left(\frac{\partial y}{\partial t}, \phi\right) + a(y, \phi) = (g, \phi) \quad \text{for all} \quad \phi \in V, \tag{1.1}$$

where we have set

$$g = f + Bv, \tag{1.2}$$

with

$$y(0) = y_0. \tag{1.3}$$

For the approximation of y we have a (large) number of different possibilities; in the *space variables* we can consider:

(i) *internal* approximations,

(ii) *external* approximations.

In case (i) we consider finite-dimensional subspaces V_h of V:[1]

$$V_h \subset V; \tag{1.4}$$

in case (ii), the approximation will belong to a finite-dimensional space V_h which is not necessarily contained in V.

Example 1.1. $V = H_0^1(\Omega)$; $\Omega \subset \mathbb{R}^2$; h = maximum length of triangles contained in Ω as in Fig. 8 (we assume that the minimum angle of the triangulation is $\geqq \theta_0 > 0$); V_h = space of continuous functions, which are zero on the boundary of the triangulation, and which are linear in each triangle; we have (1.4).

Example 1.2. V as in Example 1.1; $h = \{h_1, h_2\}$ (see Fig. 9); V_h = space generated by the characteristic functions of the rectangles shadowed on Fig. 9. In this case V_h is *not* contained in V.

---

[1] The meaning of h is explained by the examples below.

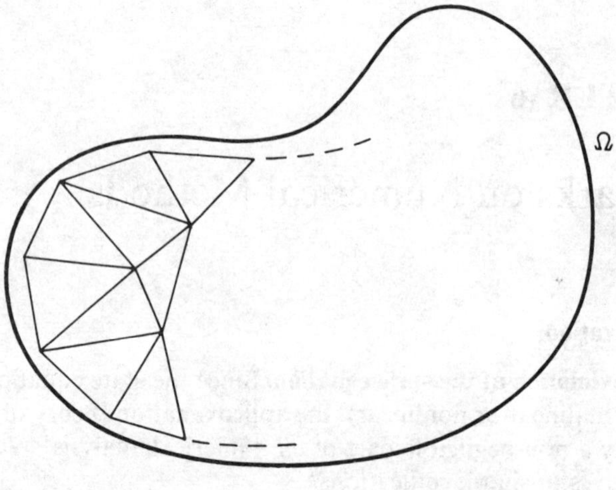

FIG. 8

A *semi-discretization* (i.e., a discretization *only* in the space variable) is immediately defined using *internal* approximations: we define $y_h(t)$ as the solution of

$$
\begin{aligned}
&\left(\frac{\partial y_h}{\partial t}, \phi_h\right) + a(y_h, \phi_h) = (g, \phi_h) \quad \text{for all} \quad \phi_h \in V_h,\\
&y_h \in V_h, \qquad\qquad (1.5)\\
&y_h(0) = y_{0h}, \quad y_{0h} \text{ “approximation” of } y_0, \quad y_{0h} \in V_h.
\end{aligned}
$$

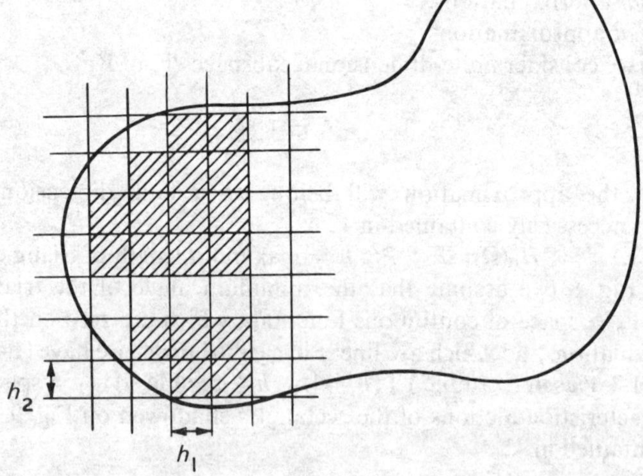

FIG. 9

The equations (1.5) are a system of *ordinary* differential equations. For the *external* approximation (1.5) does not make sense; one has then to define an "approximation" $a_h(u_h, v_h)$ of $a(u, v)$ defined on V_h;[2] the system analogous to (1.5) is then

$$
\text{(1.6)} \qquad \begin{aligned} &\left(\frac{\partial y_h}{\partial t}, \phi_h\right) + a_h(y_h, \phi_h) = (g, \phi_h) \quad \text{for all} \quad \phi_h \in V_h{}^{3} \\ &y_h(0) = y_{0h}. \end{aligned}
$$

The next step is to *discretize* in t. Let us denote

$$
\text{(1.7)} \qquad y_h^n = \text{approximation of } y_h \text{ at time } n\Delta t.^4
$$

Then, if ∇ is a finite difference approximation of $\partial/\partial t$ (to be chosen below), one approximates (1.5)[5] by

$$
\text{(1.8)} \qquad \begin{aligned} &(\nabla y_h^n, \phi_h) + a(y_h^n, \phi_h) = (g^n, \phi_h) \quad \text{for all} \quad \phi_h \in V_h, \\ &y_h^0 = y_{0h}. \end{aligned}
$$

There are very many possible choices of ∇. From a *conceptual* point of view, the two simplest choices are

$$
\text{(1.9)} \qquad \nabla y_h^n = \frac{1}{\Delta t}(y_h^n - y_h^{n-1}),
$$

$$
\text{(1.10)} \qquad \nabla y_h^n = \frac{1}{\Delta t}(y_h^{n+1} - y_h^n).
$$

The choice (1.9) leads to "implicit" methods which are "stable and convergent".[6] The choice (1.10) leads to "explicit" methods; they are "stable" and "convergent" if and only if Δt and h are related by a "stability condition"; for instance if A is a second order operator, the stability condition reads

$$
\Delta t \leqq c|h|^2.
$$

Remark 1.1. Internal and external methods were introduced by Céa [1] (cf. also Lions [13]). The internal methods lead to the so-called "finite element" methods (a trivial example being given in Example 1.1 above); we refer to Ph. Ciarlet and P. A. Raviart [1], B. Fraejs de Veubeque [1], G. Strang [1], O. C. Zienkiewicz [1], M. Zlamal [1], [2] and the bibliographies therein. For a systematic study of approximation spaces we refer to J. P. Aubin [1], [2].

Remark 1.2. For extensions of the above considerations to hyperbolic systems, Petrowski's systems and classes of nonlinear P.D.E.'s, cf. Raviart [1], Temam [2] and the bibliographies therein.

---

[2] This is done by replacing in $a(\phi, \psi)$ derivatives by finite differences.

[3] Assuming $V_h \subset H$ and g with values in H.

[4] We take *fixed* time increments; but we could also take *variable* time increments.

[5] The extension to (1.6) is straightforward.

[6] For precisions on this, consult P. A. Raviart [1], R. Temam [2].

1.2. Approximation of the optimal control problems. We now consider the following optimal control problem: we replace g in (1.1) by its value (1.2) and we want to minimize

$$J(v) = \int_0^T \|Cy(v) - z_d\|_H^2\, dt + \int_0^T (Nv, v)_{\mathcal{U}}\, dt, \tag{1.11}$$

when v spans $U_{\text{ad}} \subset U$.

We introduce a family U_ξ of *subspaces* of U:[7]

$$U_\xi \subset U, \tag{1.12}$$

where ξ denotes a "discretization" parameter and we construct a family $U_{\xi,\text{ad}}$ of closed convex subsets of U_ξ which "approximate" U_{ad}.

The *approximate state* $\{y^n_{h,\xi}\}$ is then given by[8]

$$\begin{aligned}&(\nabla y^n_{h,\xi}, \phi_h) + a(y^n_{h,\xi}, \phi_h) = (f^n + Bv^n_\xi, \phi_h) \quad \text{for all} \quad \phi_h \in V_h,\\ &y^0_{h,\xi} = y_{0h},\end{aligned} \tag{1.13}$$

where ∇ is replaced by (1.9) *or* (1.10).

The *approximate cost function* is now given by

$$J^{\Delta t}_{h,\xi}(v_\xi) = \sum_{n=0}^{M-1} \Delta t \|Cy^n_{h,\xi} - z^n_d\|_H^2 + \sum_{n=0}^{M-1} \Delta t (Nv^n_\xi, v^n_\xi)_{\mathcal{U}}, \tag{1.14}$$

where M is such that $M\Delta t = T$.

The approximate optimal control problem consists in minimizing $J^{\Delta t}_{h,\xi}(v_\xi)$ over $U_{\xi,\text{ad}}$.

Under suitable "consistency" hypotheses[9]—and, in case ∇ is chosen by (1.10), *under stability hypotheses*—we can prove (cf. Bensoussan, Bossavit and Nedelec [1]) that, if $u^{\Delta t}_{h,\xi}$ denotes the optimal control of the approximate problem, then

$$u^{\Delta t}_{h,\xi} \to u \quad \text{in } U, \tag{1.15}$$

and $y^n_{h,\xi}$ (extended as a step function) converges to y in $L^2(0, T; V)$, where u denotes the optimal control of the initial problem and where $y = y(u)$.

Remark 1.3. For other general results along these lines, we refer to Bosarge and Johnson [1], Bosarge, Johnson, McKnight and Timlake [1], Bosarge, Johnson and Smith [1] where one can find error estimates in the cases without constraints.

Remark 1.4. Internal methods are also known under the terminology of "Galerkin method".

Remark 1.5. We can extend the preceding results to nonlinear systems.

---

[7] It suffices in general to consider *internal* approximations of U—but one could very well consider also external approximations of U.

[8] There is a straightforward extension to the case of external approximation (1.6).

[9] Which express in precise terms that V_h, $U_{\xi,\text{ad}}$ are "approximations" of V, U_{ad}.

1.3. Case without constraints. In the case without constraints we can go further and, in particular, deduce from the preceding considerations, convergent approximation schemes for the solution of the nonlinear (Riccati type) P.D.E. met in Chap. 2. Results in this direction are given in Lions [1]. We explain the method—and we shall obtain some new results in the process—in the case when the control depends only on v (cf. Chap. 2, § 7[10]), i.e.,

$$\frac{\partial y(v)}{\partial t} + Ay(v) = f + v(t)w, \tag{1.16}$$

w given in V satisfying

$$Aw = \lambda w, \tag{1.17}$$

and the cost function being given by

$$J(v) = \int_0^T \|Cy(v) - z_d\|_H^2\, dt + N \int_0^T v(t)^2\, dt, \quad N > 0. \tag{1.18}$$

As we saw in Chap. 2, § 7, the optimality system is given by

$$\begin{aligned} &\frac{\partial y}{\partial t} + Ay + \frac{1}{N}(p, w)w = f, \\ &-\frac{\partial p}{\partial t} + A^*p - C^*Cy = -C^*z_d, \\ &y(0) = y_0, \quad p(T) = 0. \end{aligned} \tag{1.19}$$

We can uncouple (1.19) by

$$p = Py + r \tag{1.20}$$

and it is enough to compute $Q = Pw$, Q being characterized by

$$\begin{aligned} &-\frac{\partial Q}{\partial t} + (A^* + \lambda)Q + \frac{1}{N}(Q, w)Q = C^*Cw, \\ &Q \in L^2(0, T; V), \quad (Q, w) \geqq 0, \\ &Q(T) = 0. \end{aligned} \tag{1.21}$$

We are now going to give approximation schemes for the solution Q of (1.21). We define the appropriate state by semi-discretization:

$$\begin{aligned} &\left(\frac{\partial y_h}{\partial t}, \phi_h\right) + a(y_h, \phi_h) = (f + v(t)w_h, \phi_h) \quad \text{for all} \quad \phi_h \in V_h, \quad V_h \subset V, \\ &y_h(0) = y_0; \end{aligned} \tag{1.22}$$

---

[10] The considerations which follow readily extend to the situations of Chap. 2, §§ 8.2 and 9.

(we do not make approximations on U); in (1.22) w_h is an approximate eigenfunction, i.e.,

$$a(w_h, \phi_h) = \lambda_h(w_h, \phi_h), \quad w_h \in V_h, \quad \text{for all} \quad \phi_h \in V_h, \tag{1.23}$$

where $\lambda_h \to \lambda$ and $w_h \to w$ in V as $h \to 0$.

The approximate cost function is given by

$$J_h(v) = \int_0^T \|Cy_h(v) - z_d\|_H^2 \, dt + N \int_0^T v(t)^2 \, dt. \tag{1.24}$$

One shows easily the result analogous to (1.15). If u_h denotes the optimal control corresponding to (1.22), (1.24), the optimality system is given by

$$\begin{aligned} &\left(\frac{\partial y_h}{\partial t}, \phi_h\right) + a(y_h, \phi_h) + \frac{1}{N}(p_h, w_h)(w_h, \phi_h) = (f, \phi_h), \\ &-\left(\frac{\partial p_h}{\partial t}, \phi_h\right) + a^*(p_h, \phi_h) - (C^*Cy_h, \phi_h) = -(C^*z_d, \phi_h) \quad \text{for all} \quad \phi_h \in V_h; \\ &y_h, p_h \in L^2(0, T; V_h), \\ &y_h(0) = y_{0h}, \quad p_h(T) = 0, \end{aligned} \tag{1.25}$$

and

$$u_h = -\frac{1}{N}(p_h, w_h). \tag{1.26}$$

The uncoupling of (1.25) leads to[11]

$$p_h = P_h y_h + r_h, \tag{1.27}$$

and $(P_h\phi_h, \psi_h) = (\phi_h, P_h\psi_h)$ for all $\phi_h, \psi_h \in V_h$. Then

$$u_h = -\frac{1}{N}(P_h w_h, y_h) - \frac{1}{N}(r_h, w_h) \tag{1.28}$$

and it suffices to compute $P_h w_h$ and r_h. Let us study the computation of $P_h w_h$. If we set

$$P_h w_h = Q_h, \tag{1.29}$$

the usual identification computation leads to

$$\begin{aligned} &-\left(\frac{\partial Q_h}{\partial t}, \phi_h\right) + a^*(Q_h, \phi_h) + \lambda_h(Q_h, \phi_h) + \frac{1}{N}(Q_h, w_h)(Q_h, \phi_h) \\ &\qquad = (C^*Cw_h, \phi_h) \quad \text{for all} \quad \phi_h \in V_h, \\ &Q_h(T) = 0, \\ &Q_h \in L^2(0, T; V_h). \end{aligned} \tag{1.30}$$

---

[11] The same reasoning as in Chap. 2, §§ 6 and 7.

The convergence theorem of the approximations gives: *if* w_h *is given by* (1.23), *and if* w *satisfies* (1.17), *when* $\lambda_h \to \lambda$, *then we have*

$$Q_h \to Q \quad \text{in } L^2(0, T; V). \tag{1.31}$$

This proves the convergence of the Galerkin approximation of the nonlinear P.D.E. (1.21).

Example 1.1. We take the situation of Example 7, Chap. 2. Then the approximation Q_h of the solution Q of (7.36), Chap. 2, is given by

$$\begin{aligned} -\left(\frac{\partial Q_h}{\partial t}, \phi_h\right) + a(Q_h, \phi_h) + \frac{1}{N}\left(\int_\Omega Q_h w\, dx\right)(Q_h, \phi_h) &= \bar{w}_h \int_\Omega \phi_h\, dx \\ &\text{for all} \quad \phi_h \in V_h, \\ Q_h(x, T) = 0, \quad x \in \Omega. \end{aligned} \tag{1.32}$$

We can write in a similar manner approximations for the nonlinear P.D.E. considered in Examples 7.1 and 7.3, Chap. 2.

Remark 1.5. *Stable discretization of the nonlinear P.D.E.* (1.21). By using semi-discretization in (1.22) we obtained a convergent Galerkin approximation in (1.30). We now start from an *implicit*[12] discretization of (1.21); this will lead us to a stable (and convergent) implicit discrete approximation of (1.21).

We choose Δt and M such that $M\Delta t = T$. For simplifying the writing, we set $\Delta t = k$.

We start from

$$\begin{aligned} \frac{y^n - y^{n-1}}{k} + A_h y^n &= v^n w,^{13} \\ y^0 &= y_{0h}, \end{aligned} \tag{1.33}$$

where we write for simplicity

$$y_h^n = y^n, \quad w_h = w, \tag{1.34}$$

and where A_h is defined by

$$(A_h\phi_h, \psi_h) = a(\phi_h, \psi_h) \quad \text{for all} \quad \phi_h, \psi_h \in V_h. \tag{1.35}$$

The approximate cost function is given by[14]

$$J_h = \sum_{n=0}^{M-1} k\|Cy^n\|_H^2 + N\sum_{n=0}^{M-1} k|v^n|^2. \tag{1.36}$$

---

[12] We could also start from an *explicit* discretization; this leads to an explicit discrete approximation of (1.21), which is convergent under the stability condition.

[13] We take $f = 0$ to simplify the writing.

[14] We take $z_d = 0$ in order to simplify the exposition but it is not at all essential.

If we introduce the adjoint state by the *implicit* scheme:

$$(1.37)\qquad -\frac{p^{n+1}-p^n}{k} + A_h^* p^n = C^*Cy^n, \qquad n = M-1, M-2, \cdots, 0,$$
$$p^M = 0;$$

then the approximate optimal control u^n is given by

$$(1.38)\qquad (w, p^n) + Nu^n = 0 \quad \text{for all} \quad n.$$

The final approximate optimal system is therefore given by

$$(1.39)\qquad \frac{y^n - y^{n-1}}{k} + A_h y^n + \frac{1}{N}(p^n, w)w = 0, \qquad 1 \leqq n \leqq M,$$

$$(1.40)\qquad -\frac{p^{n+1}-p^n}{k} + A_h^* p^n - C^*Cy^n = 0, \qquad 0 \leqq n \leqq M-1,$$

$$(1.41)\qquad y^0 = y_{0h}, \quad p^M = 0.$$

The reasoning used in the continuous case for the uncoupling is still valid and leads to

$$(1.42)\qquad p^n = P^n y^n,$$

$$(1.43)\qquad P^n \in \mathscr{L}(V_h; V_h), \quad (P^n\phi_h, \psi_h) = (\phi_h, P^n\psi_h) \quad \text{for all} \quad \phi_h, \psi_h \in V_h;$$

$$(1.44)\qquad (P^n\phi_h, \phi_h) \geqq 0 \quad \text{for all} \quad \phi_h \in V_h;$$

of course P^n depends on h:

$$(1.45)\qquad P^n = P_h^n.$$

To obtain the equation satisfied by P^n, let us use (1.39) for $n+1$ instead of n, to obtain

$$(1.46)\qquad y^n = (I + kA_h)y^{n+1} + \frac{k}{N}(p^{n+1}, w)w.$$

Applying P^n to (1.46), we obtain

$$(1.47)\qquad p^n = P^n y^n = P^n(I + kA_h)y^{n+1} + \frac{k}{N}(p^{n+1}, w)P^n w.$$

On the other hand, (1.40) gives

$$p^n = (I + kA_h^*)^{-1}[p^{n+1} + kC^*Cy^n];$$

hence, using (1.46),

$$p^n = (I + kA_h^*)^{-1}[P^{n+1}y^{n+1} + kC^*C(I + kA_h)y^{n+1} + \frac{k^2}{N}C^*Cw(P^{n+1}y^{n+1}, w)].$$

Comparing (1.47) and (1.48) we obtain

$$
\begin{aligned}
&P^n(I + kA_h)y^{n+1} + \frac{k}{N}(P^{n+1}y^{n+1}, w)P^n w \\
(1.49) \qquad &= (I + kA_h^*)^{-1}[P^{n+1}y^{n+1} + KC^*C(I + kA_h)y^{n+1} \\
&+ \frac{k^2}{N}C^*Cw(P^{n+1}y^{n+1}, w)].
\end{aligned}
$$

But (1.49) is an identity in y^{n+1}. We can therefore take $y^{n+1} = w$; if we set

$$(1.50) \qquad P^n w = Q^n,$$

and if we observe that $A_h w = (A_h w_h) = \lambda_h w$ it follows that

$$
\begin{aligned}
&(1 + k\lambda_h)Q^n + \frac{k}{N}(Q^{n+1}, w)Q^n \\
(1.51) \qquad &= (I + kA_h^*)^{-1}\left[Q^{n+1} + (1 + k\lambda_h)kC^*Cw + \frac{k^2}{N}(Q^{n+1}, w)C^*Cw\right] \\
&n = M - 1, M - 2, \cdots, 0; \quad Q^M = 0.
\end{aligned}
$$

This (bad-looking) scheme can be (somewhat) cleaned up by using decomposition methods.

Let us define Q^n (assuming Q^{n+1} to be known) in three steps:

$$(1.52) \qquad -\frac{Q^{n+1} - Q^{n+2/3}}{k} + A_h^* Q^{n+2/3} = C^*Cw\left[1 + k\lambda_h + \frac{k}{N}(Q^{n+1}, w)\right],$$

$$(1.53) \qquad -\frac{Q^{n+2/3} - Q^{n+1/3}}{k} + \lambda_h Q^{n+1/3} = \frac{k\lambda_h}{N}(Q^{n+1}, w),$$

$$(1.54) \qquad -\frac{Q^{n+1/3} - Q^n}{k} + \frac{1}{N}(Q^{n+1}, w)Q^n = 0.$$[15]

By eliminating $Q^{n+1/3}, Q^{n+2/3}$ in these equations, we can see that the split-up scheme (1.52)-(1.54) is identical to (1.51).

Therefore we have proved that this scheme is a stable and convergent approximation of the solution Q of (1.21).

Remark 1.6. For other Riccati-type nonlinear P.D.E.'s, schemes of the preceding type have been introduced by J. Nedelec [1] and R. Temam [1]. For other types of P.D.E., methods of the above type are known as "splitting-up methods" or "fractional steps methods"; we refer to G. I. Marchuk [1], N. N. Yanenko [1] and the bibliographies therein, and to R. Temam [2].

2. Iterative methods. Until now in this chapter we have reduced the infinite-dimensional problem to finite-dimensional problems (of various forms), giving

[15] Let us recall that $(Q^{n+1}, w) \geqq 0$.

convergent approximations. But this is of course only the first step (and, actually, the simplest!). We have now to provide algorithms which permit the numerical solution of the finite-dimensional problems. We are now going to give some indications on this (fundamental) part of the problem.

2.1. Lagrangians. In order to clarify the discussion, we consider the general abstract situation of Chap. 2, § 1. (We shall see that the algorithms we obtain extend to nonlinear problems.) With the notation of Chap. 2, § 1, the state of the system is given by

$$\Lambda y(v) = f + Bv, \tag{2.1}$$

and the cost function by

$$J(v) = \|Cy(v) - z_d\|_H^2 + (Nv, v)_U. \tag{2.2}$$

The optimality system is given by (we denote by u the optimal control and we set $y(u) = y$):

$$\begin{aligned} &\Lambda y = f + Bu, \\ &\Lambda^* p = C^*(Cy - z_d), \\ &(Nu, v - u)_U + (B^*p, v - u)_U \geqq 0 \quad \text{for all} \quad v \in U_{\text{ad}}, \\ &u \in U_{\text{ad}}. \end{aligned} \tag{2.3}$$

Lagrangian (first form). We now introduce

$$\mathscr{L}(y, v, q) = \|Cy - z_d\|_H^2 + (Nv, v)_U - (q, \Lambda y - f - Bv), \tag{2.4}$$

where y and v are *independent*, y spans Y, v spans U_{ad}, and where q spans F' = dual space of $F(\Lambda \in \mathscr{L}(Y; F))$; $\mathscr{L}$ is called a Lagrangian.

Let us check that if $\{y, u, p\}$ is a saddle point of $\mathscr{L}$, i.e., if

$$\mathscr{L}(y, u, q) \leqq \mathscr{L}(y, u, p) \leqq \mathscr{L}(\eta, v, p) \quad \text{for all} \quad \eta, v, q, \tag{2.5}$$

then y, u, p is the solution of (2.3) and conversely.

Indeed the first inequality in (2.5) implies that

$$\Lambda y = f - Bu = 0,$$

which gives the first equation in (2.3); the second inequality in (2.5) is equivalent to

$$\begin{aligned} &\|Cy - z_d\|_H^2 - (p, \Lambda y) \leqq \|C\eta - z_d\|_H^2 - (p, \Lambda\eta) \quad \text{for all} \quad \eta, \\ &(Nu, u)_U + (p, Bu) \leqq (Nv, v)_U + (p, Bv) \quad \text{for all} \quad v \in U_{\text{ad}}. \end{aligned} \tag{2.6}$$

The first inequality in (2.6) is "equivalent"—at least formally—to

$$-(p, \Lambda\eta) + (Cy - z_d, C\eta) = 0 \quad \text{for all} \quad \eta;$$

hence $p \in D(\Lambda^*)$ and $\Lambda^* p = C^*(Cy - z_d)$. The second inequality in (2.6) is equivalent to the inequality in (2.3); hence the result follows.

There is not a unique Lagrangian having the property (2.5). Indeed, let us write (2.1) in the (obviously equivalent) form

$$y(v) = \Lambda^{-1}f + \Lambda^{-1}Bv. \tag{2.7}$$

We then introduce

$$\mathcal{M}(y, v, q) = \|Cy - z_d\|_H^2 + (Nv, v)_U - (q, y - \Lambda^{-1}f - \Lambda^{-1}Bv) \tag{2.8}$$

and one has the property analogous to (2.5):

$$\mathcal{M}(y, u, q) \leqq \mathcal{M}(y, u, p) \leqq \mathcal{M}(\eta, v, p) \quad \text{for all} \quad \eta, v, q. \tag{2.9}$$

We give now algorithms deduced from (2.5) or (2.9).

2.2. Iterative algorithms.[16] For the approximation of a saddle point, we can use, among others, a method of H. Uzawa [1].

Let us take first the Lagrangian given by (2.4). Assuming p^n to be known, we first "define" (this is *formal*) y^n, u^n by minimizing $\mathcal{L}(y, v, p^n)$, i.e.,

$$\begin{aligned} &C^*(Cy^n - z_d) = \Lambda^* p^n, \\ &(Nu^n, v - u^n)_U + (B^*p^n, v - u^n)_{U'} \geqq 0 \quad \text{for all} \quad v \in U_{\mathrm{ad}}; \end{aligned} \tag{2.10}$$

then we define p^{n+1} by

$$p^{n+1} = p^n + \rho_n \frac{\partial}{\partial q}\mathcal{L}(y^n, u^n, p^n), \qquad \rho_n > 0, \tag{2.11}$$

i.e.,

$$p^{n+1} = p^n - \rho_n(\Lambda y^n - f - Bu^n). \tag{2.12}$$

The last step (2.12) which amounts to successive application of the unbounded operator Λ is likely to lead to instabilities.

This difficulty is avoided by using the same kind of method with the Lagrangian $\mathcal{M}$ given by (2.8).

Assuming p^n to be known, we "define" y^n, u^n by minimizing $\mathcal{M}(y, v, p^n)$, i.e.,

$$\begin{aligned} &C^*(Cy^n - z_d) = p^n, \\ &(Nu^n, v - u^n)_U + (\Lambda^{*-1}p^n, B(v - u^n)) \geqq 0 \quad \text{for all} \quad v \in U_{\mathrm{ad}}. \end{aligned} \tag{2.13}$$

Then we define p^{n+1} by

$$p^{n+1} = p^n + \rho_n \frac{\partial \mathcal{M}}{\partial q}(y^n, u^n, p^n), \tag{2.14}$$

i.e.,

$$p^{n+1} = p^n - \rho_n(y^n - \Lambda^{-1}f - \Lambda^{-1}Bu^n). \tag{2.15}$$

---

[16] We express these algorithms directly on the infinite-dimensional case. Of course, in the applications, one uses these algorithms on the discretized problems as defined in § 1.

In other words, the operations to be fulfilled are:

(i) solve $C^*(Cy^n - z_d) = p^n$;
(ii) solve $\Lambda^*\pi^n = p^n$;
(iii) solve $(Nu^n + B^*\pi^n, v - u^n)_U \geqq 0$ for all $v \in U_{\text{ad}}$;
(iv) solve $\Lambda p^{n+1} = \Lambda(p^n - \delta_n y^n) + \rho_n(f + Bu^n)$.

Remark 2.1. For the preceding observations, together with numerical applications, and indications on the choice of ρ_n, cf. J. P. Yvon [2].

Remark 2.2. We can use, in a formal manner, similar methods for nonlinear systems. This has been done by J. P. Kernevez [1] for the problem of Chap. 3, § 1 and other problems arising in biochemistry.

Remark 2.3. Of course, we can apply other methods of optimization theory; cf. E. R. Barnes [1], and Cornick and Michel [1] for an application of the conjugate gradient methods. Cf. also J. Céa [2] and V. W. Daniel [1].

Remark 2.4. The algorithms given above have been extensively used for solving the unilateral problems considered in Chap. 1. We refer to D. Bégis [1], J. F. Bourgat [1], J. Céa and R. Glowinski [1], J. Céa, R. Glowinski and J. Nedélec [1], B. Courjaret [1], M. Goursat [1], I. Marrocco [1], R. Trémolières [1] and the book of Glowinski, Lions and Trémolières [1].

Remark 2.5. *Relationship to penalty methods.* We can also think of the state equation as a constraint; this leads to the introduction of a *penalty term* to take care of this constraint; we therefore introduce:

$$J_\varepsilon(y, v) = \|Cy - z_d\|_H^2 + (Nv, v)_U + \frac{1}{\varepsilon}\|\Lambda y - f - Bv\|_F^2, \qquad \varepsilon > 0. \tag{2.16}$$

It is easy to show the following: *if* $y_\varepsilon, u_\varepsilon$ *denotes the solution of* $\inf J_\varepsilon(y, v)$, $y \in Y$, $v \in U_{\text{ad}}$, *then* $y_\varepsilon \to y, u_\varepsilon \to u$ *in* Y *and* U *respectively.*

If we compare this to, say, (2.4), we see that this penalty method amounts to choosing a Lagrange multiplier q of a special form, namely,

$$q = -\frac{1}{\varepsilon}\Lambda_F(\Lambda y - f - Bv), \tag{2.17}$$

where Λ_F denotes the canonical isomorphism from $F \to F'$ in such a way that

$$-(q, \Lambda y - f - Bv) = \frac{1}{\varepsilon}\|\Lambda y - f - Bv\|_F^2. \tag{2.18}$$

The penalty method in the form (2.16) has been introduced in Lions [1] where it is shown that the method readily extends to some nonlinear systems. Other applications have been given by A. V. Balakrishnan [3], [4], together with numerical applications. Numerical applications to distributed systems have been given by J. P. Yvon [1]. Comparisons of numerical experiments with "all" possible methods for a distributed model will be presented in D. Leroy [1].

2.3. Direct methods for solving the optimality system. In the preceding sections, we solved the optimality system (2.3) by going through a Lagrangian and by using an algorithm of approximation for the saddle points.

We can also directly write iterative algorithms for solving (2.3), cf. J. C. Miellou [1], L. Tartar [2], J. P. Yvon [2]. For instance, the Gauss-Seidel algorithm gives the following scheme: assuming $\{y^n, p^n, u^n\}$ to be known, we define successively $\{y^{n+1}, p^{n+1}, u^{n+1}\}$ by

$$\Lambda y^{n+1} = f + Bu^n, \tag{2.19}$$

$$\Lambda^* p^{n+1} = C^*(Cy^{n+1} - z_d), \tag{2.20}$$

$$(Nu^{n+1}, v - y^{n+1})_U + (B^* p^{n+1}, v - u^{n+1})_U \geqq 0 \quad \text{for all} \quad v \in U_{\text{ad}}. \tag{2.21}$$

We can prove (cf. Miellou and Yvon, loc. cit.) the convergence of the method if N is "large enough", i.e.,

$$(Nv, v)_U \geqq \nu \|v\|_U^2, \quad \nu \text{ large enough}. \tag{2.22}$$

(We can indeed observe numerically (cf. Yvon [2]) instabilities when ν becomes small.)

Remark 2.6. We do not study here decomposition methods; cf. Mesarovic, Macko and Takahara [1], Bensoussan, Lions and Temam [1], and the bibliographies therein.

REFERENCES

J. K. AGGARWAL [1] *Feedback control of linear systems with distributed delay*, IFAC Symposium on the Control of Distributed Parameter Systems, Banff, June, 1971.

M. ATHANS AND P. L. FALB [1] *Optimal Control*, McGraw-Hill, New York, 1966.

J. P. AUBIN [1] *Approximation des espaces de distributions et des opérateurs différentiels*, Mémoires Société Mathématique France, 12 (1967), pp. 3–139.

——— [2] Book to appear.

C. BAIOCCHI [1] C.R. Acad. Sciences, Paris, December, 1971.

A. V. BALAKRISHNAN [1] *Optimal control problems in Banach spaces*, SIAM J. Control, 3 (1965), pp. 152–180.

——— [2] *Semi-group theory and control theory*, Proc. I.F.I.P., Washington, D.C., Spartan Books, 1965.

——— [3] *A new computing technique in system identification*, J. Computer and System Science, 2 (1968), pp. 102–116.

——— [4] *On a new computing technique in optimal control*, SIAM. J. Control, 6 (1968), pp. 149–173.

H. T. BANKS, M. Q. JACOBS AND M. R. LATINA [1] *The synthesis of optimal controls for linear problems with retarded controls*, Center for Dynamical Systems, Brown University, Providence, R.I., 1971.

E. R. BARNES [1] *Necessary and sufficient optimality conditions for a class of distributed parameter control systems*, SIAM J. Control, 9 (1971), pp. 62–82.

——— [2] *Computing optimal controls in systems with distributed parameters*, IFAC Symposium on the Control of Distributed Parameter Systems, Banff, June, 1971.

D. BEGIS [1] Thesis, 3d Cycle, Paris, 1972.

A. BENSOUSSAN [1] *Identification et filtrage*, Cahiers de l'IRIA, no. 1 (1969), pp. 1–233.

——— [2] *Filtrage Optimal des Systèmes Linéaires*, Dunod, Paris, 1971.

——— [3] *Systèmes à retard*, to appear.

——— [4] *On the separation principle for distributed parameter systems*, IFAC Symposium on the Control of Distributed Parameter Systems, Banff, June, 1971.

A. BENSOUSSAN, A. BOSSAVIT AND J. C. NEDELEC [1] *Approximation des problèmes de contrôle optimal*, Cahiers de l'IRIA, no. 2 (1970), pp. 107–176.

A. BENSOUSSAN AND J. L. LIONS [1] *On the filtering theory for some stochastic distributed parameter systems*, to appear.

A. BENSOUSSAN, J. L. LIONS AND R. TEMAM [1] Cahiers de l'IRIA, 1972.

F. BIDAUT [1] Thesis, Paris, to appear.

W. E. BOSARGE, JR. AND O. G. JOHNSON [1] *Error bounds of high order accuracy for the state regulator problem via piecewise polynomial approximation*, SIAM J. Control, 9 (1971), pp. 15–28.

W. E. BOSARGE, JR., O. G. JOHNSON, R. S. MCKNIGHT AND W. P. TIMLAKE [1] *The Ritz-Galerkin procedure for non-linear control problems*, I.B.M. Scientific Center, Houston, May, 1971.

W. E. BOSARGE, JR., O. G. JOHNSON AND C. L. SMITH [1] *A direct method approximation to the linear parabolic regulator problem over multivariate spline basis*, I.B.M. Scientific Center, Houston, December, 1970.

J. F. BOURGAT [1] *Analyse numérique du problème de la torsion élastoplastique*, Thesis, 3d Cycle, Paris, 1971.

BRAUNER AND PENEL [1] Thesis, 3d Cycle, Paris, 1972.

H. BRÉZIS [1] *Sur la régularité de la solution d'inéquations elliptiques*, Ann. Inst. Fourier, 18 (1968), pp. 115–175.

——— [2] *Inéquations variationnelles*, J. Math. Pures Appl., 51 (1972), to appear.

H. BRÉZIS AND G. STAMPACCHIA [1] *Sur la régularité de la solution d'inéquations elliptiques*, Bull. Soc. Math. France, 96 (1968), pp. 153–180.

R. S. BUCY AND P. D. JOSEPH [1] *Filtering for Stochastic Processes with Applications to Guidance*, Interscience, New York, 1968.

M. M. BURGER AND K. D. NOOMAN [1] *Restoration of normal growth by covering of agglutinin sites on tumour cell surface*, Nature, 228 (1970), pp. 512–515.

A. G. Butkovskii [1] *Theory of Optimal Control of Distributed Parameter Systems*, Moscow, 1965. (English translation: American Elsevier, 1969.)

J. Céa [1] *Approximation variationnelle des problèmes aux limites*, Ann. Inst. Fourier, 14 (1964), pp. 345–444.

——— [2] *Optimisation: Théorie et Algorithmes*, Dunod, Paris, 1970.

J. Céa and R. Glowinski [1] *Méthodes numériques pour l'écoulement laminaire d'un fluide rigide viscoplastique incompressible*, to appear.

J. Céa, R. Glowinski and J. Nedelec [1] to appear.

L. Cesari [1] *Multi-dimensional Lagrange problems of optimization in a fixed domain and an application to a problem of magneto-hydrodynamics*, Arch. Rational. Mech. Anal., 29 (2) (1968), pp. 81–104.

Ph. Ciarlet and P. A. Raviart [1] *Approximation Numérique des Solutions des Problèmes aux Limites Elliptiques*, to appear.

D. E. Cornick and A. N. Michel [1] *Numerical optimization of distributed parameter systems by gradient methods*, IFAC Symposium on the Control of Distributed Parameter Systems, Banff, June, 1971.

B. Courjaret [1] Thesis, 3d Cycle, Paris, 1972.

J. W. Daniel [1] *Approximate minimization of functionals by discretization: numerical methods in optimal control*, Center for Numerical Analysis, The University of Texas, Austin, September, 1970.

G. Da Prato [1] *Equations d'évolution dans des algèbres d'opérateurs et applications*, J. Math. Pures Appl., 48 (1969), pp. 59–107.

——— [2] *Somme d'applications non linéaires*, Rome Symposium, May, 1970.

——— [3] *Quelques résultats d'existence et régularité pour un problème non linéaire de la théorie du contrôle*, Bordeaux, May, 1971.

M. C. Delfour and S. K. Mitter [1] *Hereditary differential systems with constant delays, I, II*, C.R.M., University of Montreal, 1971.

J. Dieudonné [1] *Foundations of Modern Analysis*, Academic Press, New York, 1960.

J. Duvaut and J. L. Lions [1] *Sur les Inéquations en Mécanique et en Physique*, Dunod, Paris, 1972.

M. Edelstein [1] *On nearest points of sets in uniformly convex Banach spaces*, J. London Math. Soc., 43 (1968), pp. 375–377.

I. Ekeland [1] *Sur le controle optimal de systèmes gouvernés par des équations elliptiques*, J. Functional Analysis, 1972.

H. O. Fattorini [1] *Some remarks on complete controllability*, SIAM J. Control, 4 (1966), pp. 686–694.

G. Fichera [1] *Problemi elastostatici con vincoli unilaterali; il problema di Signorini con ambigue condizioni al contorno*, Atti. Accad. Naz. Lincei Mem Cl. Sci. Fis. Mat. Nat., 8 (7), (1964), pp. 91–140.

W. Fleming [1] *The Cauchy problem for a nonlinear first order partial differential equation*, J. Differential Equations, 5 (1969), pp. 515–530.

N. D. Fowkes [1] *A singular perturbation method, I, II*, Quart. Appl. Math., 26 (1968), pp. 57–59 and 71–85.

B. Fraejs de Veubeque [1] *Displacement and equilibrium models in the finite element method*, Stress Analysis, O. C. Zienkiewicz and G. S. Holister, eds., John Wiley, London, 1965, Chap. 9.

R. V. Gamkrelidze [1] *On some extremal problems in the theory of differential equations with applications to the theory of optimal control*, SIAM J. Control, 3 (1965), pp. 106–128.

R. Glowinski, J. L. Lions and R. Tremolieres [1] *Sur la Résolution Numérique des Inéquations de la Mécanique et de la Physique*, Dunod, Paris, 1972.

M. Goursat [1] *Analyse numérique de problèmes d'élastoplasticité et de viscoplasticité*, Thesis, 3d Cycle, Paris, 1971.

J. K. Hale [1] *Dynamical systems and stability*, J. Mathematical Applications, 26 (1969), pp. 39–59.

Ph. Hartman and G. Stampacchia [1] *On some nonlinear elliptic differential functional equations*, Acta Math., 115 (1966), pp. 271–310.

Y. Haugazeau [1] *Sur les inéquations variationnelles*, C.R. Acad. Sci., Paris, (1967).

A. D. IOFFE AND V. M. TIKOMIROV [1] *Relaxed variational problems*, Trudy Moskov. Mat. Obsc., 18 (1968), pp. 187–266.

R. E. KALMAN AND R. S. BUCY [1] *New results in linear filtering and prediction theory*, Trans. ASME Ser. D. J. Basic Engrg., 83 (1961), pp. 95–107.

J. P. KERNEVEZ [1] Thesis, Paris, 1972.

——— [2] to appear.

J. P. KERNEVEZ AND THOMAS [1] Book, in preparation.

J. KEVORKIAN [1] *The two variable expansion procedure for the approximate solution of certain nonlinear differential equations*, Lectures in Applied Mathematics, vol. 7, Part 3. Space Mathematics, J. B. Rosser, ed., American Mathematical Society, Providence, R.I., 1966, pp. 206–275.

P. V. KOKOTOVIC AND P. SANNUTI [1] *Singular perturbation method for reducing the model order in optimal control design*, IEEE Trans. Automatic Control, 13 (1968), pp. 377–834.

P. V. KOKOTOVIC AND R. A. YACKEL [1] *Singular perturbation theory of linear state regulators*, Proc. Eighth Annual Allerton Conference, Circuit and Systems Theory, 1970, pp. 310–321.

L. B. KOPPEL, Y. P. SHIH AND D. R. CONGHANOWA [1] *Optimal feedback control of a class of distributed parameter systems with space independent controls*, I. and E.C. Fundamentals, 7 (1968), pp. 286–295.

P. KRÉE [1] *Contribution to the linear theory of distributed parameters systems*, to appear.

H. J. KUSHNER [1] *On the optimal control of a system governed by a linear parabolic equation with "white noise" inputs*, SIAM J. Control, 6 (1968), pp. 596–614.

H. J. KUSHNER AND D. I. BARNEA [1] *On the control of a linear functional-differential equation with quadratic cost*, SIAM J. Control, 8 (1970), pp. 257–272.

E. B. LEE AND L. MARKUS [1] *Foundations of Optimal Control Theory*, John Wiley, New York, 1967.

D. LEROY [1] Thesis, 3d Cycle, Paris, 1972.

H. LEWY AND G. STAMPACCHIA [1] *On the regularity of a solution of a variational inequality*, Comm. Pure Appl. Math., to appear.

J. L. LIONS [1] *Contrôle Optimal de Systèmes Gouvernés par des Équations aux Dérivées Partielles*, Dunod, Gauthier Villars, Paris, 1968. (English translation by S. K. Mitter, Grundlehren, Springer 170, 1971.)

——— [2] *Quelques Méthodes de Résolution des Problèmes aux Limites non Linéaires*, Dunod, Gauthier Villars, Paris, 1969. (English translation by Le Van, Holt, Rinehart, and Winston, New York, 1972.)

——— [3] *Sur le contrôle optimal de systèmes décrits par des équations aux dérivées partielles linéaires. (I) Equations elliptiques, (II) Equations d'évolution*, C.R. Acad. Sci. Paris, 263 (1966), pp. 713–715; 776–779.

——— [4] *Optimisation pour certaines classes d'équations d'évolution non linéaires*, Ann. Mat. Pura Appl., LXXII (1966), pp. 275–294.

——— [5] *Sur quelques problèmes d'optimisation dans les équations d'évolution linéaires de type parabolique*, Applications of Functional Analysis to Optimisation.

——— [6] *On some nonlinear partial differential equations related to optimal control theory*, Proc. Symposium Pure Mathematics, XVIII, Part 1, Chicago, 1968, A.M.S. Publication 1970, pp. 169–181.

——— [7] *Equations Différentielles Opérationnelles et Problèmes aux Limités*, Springer, Berlin, 1961.

——— [8] *On partial differential inequalities*, Uspekhi Mat. Nauk, 26:2 (1971), pp. 206–263. (In Russian.)

——— [9] *Quelques résultats d'existence dans les équations aux dérivées partielles non linéaires*, Bull. Soc. Math. France, 87 (1959), pp. 245–273.

——— [10] *Singular perturbations and singular layers in variational inequalities*, Symposium on Nonlinear Functional Analysis. Madison, Wisconsin, April, 1971.

——— [11] *Sur les perturbations singulières et les développements asymptotiques dans les équations aux dérivées partielles*, C.R. Acad. Sci., Paris, 272 (1971), pp. 995–998.

——— [12] *Perturbations singulières et couches limites dans les problèmes aux limites et le contrôle optimal*, Leray Seminar, Collège de France, December 1971.

——— [13] *Sur l'approximation des solutions de certains problèmes aux limites*, Rend. Sem. Pat. Padova, XXXII (1962), pp. 3–54.

——— [14] *Optimal control of Deterministic Distributed Parameter Systems*, IFAC Symposium on the Control of Distributed Parameter Systems, Banff, 1971, to appear in Automatica.

J. L. LIONS AND E. MAGENES [1] *Problèmes aux Limites Non Homogènes et Applications*, vol. 1, 2, 3, Dunod, Paris, 1968, 1970. (English translation by P. Kenneth, Springer, 1971, 1972.)

J. L. LIONS AND G. STAMPACCHIA [1] *Variational Inequalities*, Comm. Pure Appl. Math., XX (1967), pp. 493–519.

K. A. LURÉ [1] *Optimal control of conductivity of a fluid moving in a channel in a magnetic field*, P.M.M., 28 (1964), pp. 258–267.

G. I. MARCHOUK [1] *Numerical Methods for Meteorological Forecast*, Leningrad, 1967 (in Russian); French translation, A. Colin, Paris, 1969.

I. MARROCCO [1] Thesis, 3d. Cycle, Paris, 1970.

E. J. MCSHANE [1] *Optimal controls, relaxed and ordinary*, Mathematical Theory of Control, A. V. Balakrishnan and L. W. Neustadt, eds., Academic Press, New York, 1967, pp. 1–9.

M. D. MESARLOVIC, D. MACKO AND Y. TAKAHARA [1] *Theory of Hierarchical Multilevel Systems*, Academic Press, New York, 1970.

J. C. MIELLOU [1] Thesis, Grenoble, 1970.

S. K. MITTER [1] *Optimal control of distributed parameter systems*, Control of Distributed Parameter Systems, 1969, J.A.C.C., Boulder, Colorado, 1969, pp. 13–48.

J. C. NEDELEC [1] *Schémas d'approximations pour des équations intégro-différentielles de Riccati*, Thesis, Paris, 1970.

L. NIRENBERG [1] *Remarks on strongly elliptic partial differential equations*, Comm. Pure Appl. Math., 8 (1955), pp. 648–674.

R. E. O'MALLEY, JR. [1] *The singular perturbed linear state regulator problem*, to appear.

——— [2] *Singular perturbation of the time invariant linear state regulator problem*, to appear.

E. PARDOUX [1] *Feedback des systèmes à retard*, to appear.

L. S. PONTRYAGIN, V. B. BOLTYANSKII, R. V. GAMKRELIDZE AND E. F. MISCHENKO [1] *The Mathematical Theory of Optimal Processes*, Interscience, New York, 1962.

W. PRAGER AND P. G. HODGE [1] *Theory of Perfectly Plastic Solids*, John Wiley, New York, 1961.

P. A. RAVIART [1] *Sur l'approximation de certaines équations d'évolution linéaires et non linéaires*, J. Math. Pures Appl., 46 (1967), pp. 11–107; 109–183.

P. J. REEVE [1] *Optimal control for systems which include pure delays*, Intern. J. Control, 11 (1970), pp. 659–681.

R. T. ROCKAFELLAR [1] *Convex Analysis*, Princeton University Press, Princeton, N.J., 1969.

A. C. ROBINSON [1] *A survey of optimal control of distributed parameter systems*, Rep. 69-0171. Aerospace Research Laboratory, November, 1969.

D. L. RUSSELL [1] *Optimal regulation of linear symmetric hyperbolic systems with finite-dimensional controls*, J. SIAM Control, 4 (1966), pp. 276–294.

——— [2] *On boundary value control of linear symmetric hyperbolic systems*, Mathematical Theory of Control, A. V. Balakrishnan and L. W. Neustadt, eds., Academic Press, New York, 1967, pp. 312–321.

——— [3] *Linear stabilization of the linear oscillator in Hilbert space*, J. Math. Anal. Appl., 3 (1969), pp. 663–675.

——— [4] *Boundary value control of the higher dimensional wave equation*, SIAM J. Control, 9 (1971), pp. 29–42.

——— [5] *Control theory of hyperbolic equations related to certain questions in harmonic analysis and spectral theory*, to appear.

P. SANNUTI AND P. V. KOKOTOVIC [1] *Near optimum design of linear systems by a singular perturbation method*, IEEE Trans. Automatic Control, 14 (1969), pp. 15–22.

——— [2] *Singular perturbation method for near optimum design of high order nonlinear systems*, Automatica, 5 (1969), pp. 773–779.

L. SCHWARTZ [1] *Théorie des noyaux*, Proc. International Congress of Mathematics, 1 (1950), pp. 220–230.

——— [2] *Théorie des Distributions*, vols. 1, 2, Hermann, Paris, 1950, 1951.

A. SIGNORINI [1] *Sopra alcune questioni di Elastostatica*, Atti della Società Italiana per il Progresso delle Scienze, 1933.

S. L. SOBOLEV [1] *Applications of Functional Analysis to Mathematical Physics*, Leningrad, 1950. (In Russian.)

G. STRANG [1] Lectures at I.R.I.A., Paris, 1971.

L. TARTAR [1] Unpublished remark.

——— [2] *Sur les M-matrices*, to appear.

R. TEMAM [1] *Sur l'équation de Riccati associée à des opérateurs non bornés, en dimension infinie*, J. Functional Analysis, 7 (1971), pp. 85–115.

——— [2] *Sur la stabilité et la convergence de la méthode des pas fractionnaires*, Ann. Mat. Pura Appl., IV (79) (1968), pp. 191–380.

R. TREMOLIERES [1] Thesis, Paris, 1972.

H. UZAWA [1] *Studies in Linear and Nonlinear Programming*, K. J. Arrow, L. Hurwicz and H. Uzawa, eds., Stanford University Press, 1958.

M. I. VISIK AND L. A. LYUSTERNIK [1] *Regular degeneration and boundary layer for linear differential equations with small parameter*, Uspekhi Mat. Nauk, 12 (1957), pp. 3–122; Amer. Math. Soc. Trans. Ser., 2, 20 (1962), pp. 239–364.

J. WARGA [1] *Relaxed variational problems*, J. Math. Anal. Appl., 4 (1962), pp. 111–128.

N. M. YANENKO [1] *Méthode à Pas Fractionnaires*, A. Colin, Paris, 1968. (Translated from the Russian.)

L. C. YOUNG [1] *Generalized curves and the existence of an attained absolute minimum in the calculus of variations*, C.R. Société Sciences et Lettres Varsovie, 30 (1937), pp. 212–234.

J. P. YVON [1] *Application de la pénalisation à la résolution d'un problème de contrôle optimal*, Cahiers de l'IRIA, no. 2 (1970), pp. 4–46.

——— [2] *Applications des méthodes duales au contrôle optimal*, to appear.

——— [3] *Contrôle optimal de systèmes distribués à multi-critères*, to appear.

O. C. ZIENKIEWICZ [1] *The Finite Element Method in Structural and Continuum Mechanics*, McGraw-Hill, London, 1967.

M. ZLAMAL [1] *On the finite element method*, Numer. Math., 12 (1968), pp. 394–409.

——— [2] *On some finite element procedures for solving second order boundary value problems*, Numer. Math., 14 (1969), pp. 42–48.

Added in proof. In relation to Chap. 3, § 4, results on the relaxation have been obtained in:

H. BERLIOCCHI AND J. M. LASRY, *Sur le controle optimale de systèmes gouvernés par des équations aux dérivées partielles*, C.R. Acad. Sci. Paris, January, 1972.

An extension of Edelstein's theorem has been given by:

J. BARANGER, *Existence de solutions pour des problèmes d'optimisation non convexe*, C.R. Acad. Sci. Paris, February, 1972.

In addition counterexamples have been obtained by:

F. MURAT [1] *Un contre exemple pour le problème de contrôle dans les coefficients*, C.R. Acad. Sci. Paris, October, 1971.

——— [2] *Théorèmes de non existence pour des problèmes de contrôle dans les coefficients*, Ibid., February, 1972.

In relationship to Chap. 6, § 2, we wish to mention:

J. C. MIELLOU [1] *Méthode de l'état adjoint par relaxation*, A.F.C.E.T., 1972.

Also related work by A. DI GUGLIELMO, to appear.

In connection with Remark 2.6 one should add the following book:

D. A. WISMER, Editor, *Optimization Methods for Large Scale Systems with Applications*, McGraw-Hill, New York, 1971.